NEW TECHNOLOGIES, SHARED FACILITIES AND THE INNOVATORY FIRM

New Technologies, Shared Facilities and the Innovatory Firm

DEREK BOSWORTH
CHRIS JACOBS
JACKIE LEWIS

Institute for Employment Research
University of Warwick

Avebury

Aldershot · Brookfield USA · Hong Kong · Singapore · Sydney

Published by
Avebury
Gower Publishing Company Limited
Gower House
Croft Road
Aldershot
Hants GU11 3HR
England

Gower Publishing Company
Old Post Road
Brookfield
Vermont 05036
USA

Printed and Bound in Great Britain by
Athenaeum Press Ltd., Newcastle upon Tyne.

ISBN   0 566 07157 6

Contents

List of tables and figures

Figures

Figures

Preface and acknowledgements

This book reports the results of the programme of work undertaken on the ESRC project, 'Shared Facilities, Human Capital and Growth in Innovatory Small Enterprises', supported by pre-project funding under the 'New Technologies and the Firm' initiative. The research funding was given to Derek Bosworth (Institute for Employment Research, University of Warwick) and Jackie Lewis (School of Economics and Social Studies, The Polytechnic, Wolverhampton and Associate Fellow of the IER). Chris Jacobs was appointed to work on the project, in particular to assist with the library search and to undertake the main interviews. The ESRC work has lead on to further research in this area some of which is also reported. The results have contributed to two papers on barriers to growth in small innovatory firms. In addition, the book also reports on work undertaken for Leicester City Council on the possible costs and benefits of a CAD/CAM resource centre for local textiles firms.

The authors wish to acknowledge the cooperation of many individuals who have contributed to the preliminary telephone search, case study, and user-firm survey information which was collected in the course of the ESRC project. In particular, we wish to thank the following for their patience and willing cooperation during the course of this research: (case study project facilities) Mr Graham Bayley of the 'Turnkey' Project, Walsall College of Technology; Mr Clive E. Evans of the Telford ITEC, Telford; Ms Carol Hayden of the West Midlands Enterprise Board, Clothing Resource Centre Ltd., Smethwick; Mr Ray McCafferty and Mr L. John Yates of the CADCAM Microsystems Centre, The Polytechnic of Wolverhampton; Mr Neil Marsh of the Open Access Centre, Dudley College of Technology; (user-firm survey firms) Mr Brian Chapman, Drawing Office Supervisor of Sandvik Ltd.; Mr Roy Evans, Training Manager of GKN Sankey Ltd.; Mrs Ursula M. Griffiths, Clifford Williams & Son Ltd.; Mr Alan Hodson, Personnel Manger of Link 51; Mr John Thomas, Production Director of Anyspeed Ltd.

We would also like to express our thanks to the personnel in the Leicester based firms who were kind enough to complete and return our questionnaire carried out as part of the project funded by Leicester City Council. Without their cooperation the study could not have been undertaken. We are especially grateful to those companies on which the case studies were centred. We were made extremely welcome at their premises and gained great insights into the industry from the very detailed information and opinions that they so freely gave. Additionally we would like to thank all those associated with the industry or with some knowledge of the industry or its technology who were able and willing to help us with our research. Finally we would like to acknowledge the collaboration and thank the representatives of the authority, Alistair Reid, Mark Froud and Mick Roberts, who were then working for Leicester City Council, for their constructive advice throughout the project.

Needless to say, the views expressed here are those of the authors who also accept responsibility for any remaining errors.

Summary

Background

The small firm sector has grown significantly in size in recent years. However, increasing concern has been expressed recently about the ability of small firms to undertake both technological and organizational innovation, enabling development and growth to take place.

The report focuses primarily on the Midlands, within which the majority of the detailed case studies were carried out. Overall, at least until very recent times, the picture has been one of relative decline both in terms of the national and international position of firms in the region. The region was heavily reliant on manufacturing and particularly on textiles in the East and engineering in the West.

There is evidence that, in general, firms in the region have been relatively slow to respond to the challenge of our main industrial competitors. In particular, there is growing evidence of under-achievement in the assimilation of new technologies, particularly micro-electronics. This is closely linked with the relatively low levels of investment in human capital. It is argued that the regeneration of firms within the region depends heavily on raising the level of technological capability, and the level of skills in the work force.

Much of this discussion applies across firms independent of their size. However, it is fairly well established that small firms are slower to innovate than their larger counterparts. In addition, they face the added complication of making the transition from small- to large-firm management and organization. One route of potentially special relevance to small firms is the provision of shared facilities.

Barriers to growth and the coexistence of small and large firms

Whether shared facilities are able to help small firms develop depends essentially on what the barriers to growth are among small firms. Why do small firms not adopt more advanced technologies, become more competitive and grow? The report begins to tease out some of the influences at work in this conundrum by examining some of the findings of a recent ACARD project on barriers to growth, and by developing an analytical framework within which small and large firms are both represented.

The stylised facts appear to be that small firms coexist alongside larger firms, using less advanced technologies and exhibiting lower growth potential. Smaller firms have a limited capacity to generate funds internally (i.e. retained profits), face higher rates of interest on external finance, and pay lower wages. In practice, lower wages appear to be achieved at the cost of a lower quality of work force.

The report attempts to develop an analytical framework which can be used to begin to piece together the main parts of this jigsaw puzzle. First, why is it that low technology/low skill (often small) firms can coexist alongside high technology/high skill (often large) firms in the same industry? Second, why is it that smaller firms find it more difficult to grow? Third, how do changes in technology, skills and organization mesh together?

The framework used is an adaptation of the Johansen production schema, segmented by size of firm. Small firms are faced by relatively lower wages and relatively higher per unit costs of capital. The result is that small firms can coexist with large, equally profitably, using less capital intensive production methods.

The price of this coexistence is that, generally, smaller firms have less experience in more advanced technology areas (often developed by larger firms and more appropriate to their factor price ratios). On average, they have lower skill levels and therefore are less able and less likely to be successful in innovation associated with both technological and organizational change.

Finally, it is shown that as some price shock, caused for example by increased competition, is pushed through the system, there are essentially two types of response that can restore the firm's competitive position. The first is to introduce improved technologies which raise product quality or increase factor productivity. The second is to maintain the existing product and process technologies, but to seek lower cost sources of inputs.

It is argued that, given their existing levels of technology and potential for change, small and large firms will tend to choose different routes. Large firms will attempt to restore their competitive position by improving their product and/or process technologies. Small firms will tend to maintain their technology but seek cheaper inputs. The latter will often be at the cost of a lower quality of inputs, which further reduces the firm's future ability to respond to later competitive shocks by changing their technology, because the ability to innovate successfully depends on the quality of the work force.

On balance, if this hypothesis is correct, small UK firms have tended to follow a largely dead-end route because the scope for cost reductions of this type are limited, while the scope for technological improvements are open ended. A vicious circle of this type appears to require the injection of knowledge and skills necessary to achieve the

transition from low to high technology. One way of achieving this is
via shared facilities.

Textiles and clothing industry

The textiles and clothing industry in Leicester has experienced steady
long term declines in employment with textiles losing jobs at a faster
rate. The clothing industry is likely to be greater than any official
statistics would indicate since it is thought that the 'black economy'
is active in this industry. The industry has a large number of small
production units, with 60-70 per cent of the work force employed in
units of fewer than 50 employees in both clothing and textiles
production and less than 5 per cent of the work force employed in
units of over 1,000 employees. The industry in Leicester is dominated
by knitting companies. There are thought to be between 500 and 600
knitting companies in Leicester.

Output in clothing and textiles grew slowly in the 1950s, 1960s and
early 1970s, then declined up to 1982, with sharp falls in 1974 and
1980. There has been some growth since 1982 but output growth has
recently stagnated.

In the early 1980s prospects for the industry looked doomed with
dampened sales in the home market and a high value of the pound
inhibiting exports. Since then domestic consumer expenditure has been
rising rapidly, though the share of that expenditure on textiles,
clothing and footwear has fallen. Nevertheless, some sectors of the
high quality markets have been very prosperous and some of the
Leicester firms have been among those succeeding in these markets.
This was highlighted when many of the case study companies reported
their difficulties in meeting their customer requirements. The growth
in capital purchases in recent years also reflects optimism concerning
future prospects.

Technical change in the industry has been somewhat erratic with a
consequent variation in the growth of productivity in both clothing
and textiles. Between the mid-1950s and mid-1970s productivity grew
by an average 2.5 per cent per annum, followed by little change until
a drop in 1980/81; this was followed by an average rate of growth of
productivity of 5 per cent in the early 1980s.

The clothing industry is extremely competitive both domestically and
internationally. Competition comes from both low and high wage
economies, with low wage countries producing long runs and therefore
competing on price, whilst higher wage economies concentrate on
quality, design or specialist markets where shorter runs are the norm
and margins are higher. The textile industry consists of slightly
larger manufacturing units. Longer production runs are the norm in
textile production. Competition from abroad, based partly on
relatively low labour costs, is frequently exacerbated because of
government subsidies to capital purchases. Consequently, both the
clothing and textile industries have experienced a faster growth in
the import penetration ratio than the export sales ratio, with a
widening of the trade gap and a fall in the UK share of the world
market.

Increasing demands for a wider variety of materials and garments in
an industry where fashion changes, always frequent, but now perpetual,
cause problems for producers. Firms are facing increasing demands by
their customers to respond quickly to orders, particularly from large
chain stores. The Japanese-style 'just-in-time' delivery requirement

where customers place orders only days before they require delivery and expect the supplier to carry the costs of labelling, packing, storing and delivery, places additional burdens on the producers.

Estimates of employment in clothing and textiles in Leicester are between 17,000 and 25,000, with approximately 70 per cent of the work force female. The industry is a low paid industry relative to other manufacturing industries and has a 'sweatshop' image. This image is not altogether unfounded with many firms operating in dilapidated premises where noise levels are high, work loads heavy and take-home pay minimal. Nevertheless, there are also more positive aspects of the industry with many firms operating in more glamorous markets where production is more specialist and involves shorter production runs and more hand finishing. Similarly, the industry offers some career jobs in management, sales, design and so on. Also in some firms the nature of the work has been substantially changed by the introduction of new micro-electronic production technologies, making some jobs much more technical and less monotonous.

In Leicester the knitting manufacturers have moved most rapidly into the new technological age. This is perhaps because the production of knitwear has offered most scope for advance. Computer aided design (CAD) systems enable the design of garments and knitted fabric to be undertaken using a micro-computer, then, simple changes in design and various sizes of either design (e.g. a manufacturer may be asked to produce a logo in different sizes on coordinated products) or garments can be made. Computer-based machinery has been created to increase the speed at which the thread is knitted. The link between the CAD system and the machinery can also take place through a micro-computer system (perhaps the same machine as the one on which the design was generated) to 'instruct' the machinery. The production of clothing however still remains labour intensive involving an operator and stitching machine to complete the garments.

Results of Leicester survey and case study interviews

The survey drew responses from a wide variety of firms producing a range of products representing both textile and clothing firms and many of the subsectors within these broad categorizations. Knitwear and hosiery firms were well represented.

The survey revealed a preponderance of relatively young firms, although there were some very old established firms. This suggests that while the industry is long established, it suffers from a high turnover of companies with a high start-up rate but also a high failure rate.

The majority of firms have limited liability status and are single manufacturing units managed by the owner. Around a third of the companies were owned and managed by Asians while the remaining two-thirds were white owned or managed. Asian owned firms employ a majority of Asian workers while white owned firms employ a majority of white workers. Most firms, however, employ both white and Asian workers.

A large proportion of firms are small; most firms employ outworkers and about half employ freelance specialists and a similar proportion sub-contract work from time to time. About 70 per cent of the work force is female.

The major source of finance used by firms is banks and very few firms had experienced problems in raising finance. It must however be

said that we are inevitably only able to contact firms who survive the
difficulty in the first few years. Firms who have difficulty raising
finance are less likely to have survived and therefore are unlikely to
be located by surveys such as ours.

Just over half the firms had increased their turnover in the year
1986-87; 40 per cent reported stability; only a minority had therefore
suffered a decline. Almost 70 per cent of firms anticipated an
improvement in their trading position over the next 12 months at the
time of the survey.

A large proportion of firms (44 per cent) produce their own fabric
which they then cut and sew. Many firms buy in fabric to cut and sew;
smaller numbers produce their own knitted fabric to sell on or produce
hosiery or body blanks. Other activities include dying and finishing,
trimming, importing and processing raw yarn through to finished
products.

The single most important outlet for the firms was wholesalers
followed by large chain stores, who were used by just under half of
the firms. Single retail outlets and mail order companies were also
important outlets for firms. Few firms use factory shops as an
outlet. Nearly all firms sell some of their output in Britain outside
the East Midlands. A majority also sell locally and just under a
half export some of their output. The EC is the most common export
destination, with Eire being the main customer within the EC.

Only a small proportion of firms use entirely British-made fabric or
yarn, though this is also true of foreign-produced materials since
the majority of firms use a combination of foreign and British-made
materials. A high proportion of firms in Leicester use local
suppliers of yarn or fabric. There is obviously therefore some scope
for influencing the sourcing policies of firms by collaborating
with wholesalers of yarn and fabric (e.g. to identify reasons for
purchasing from abroad and problems with local and British supplies).

A number of problems were identified in the case study interviews:
fierce competition from both the home and international market; staff
recruitment problems of both skilled and semi-skilled workers (it was
clear from the interviewees' comments that the term 'skill' was being
used to cover a range of personal characteristics such as commitment,
enthusiasm, the ability to accept flexible and heavy workloads, in
addition to abilities to undertake tasks which are included in the
usual definition of skill); demands from customers for quickly
changing products and colour matching; and difficulties associated
with the acquisition of materials of suitable quality.

Companies in Leicester trade with a whole range of customers from
the market stall end of the market to the West End stores. The case
study firms reported a range of conditions and standards imposed by
customers on production techniques and processes. In particular,
firms reported the domination of large retail outlets on almost every
aspect of production from sourcing through the production procedures,
labelling, packing and delivery of goods resulting in a loss of
autonomy by the manufacturers. Some companies had decided not to
compete for orders in this area because of the constraints and
potential over-reliance on these orders which might ensue.

Competitiveness and technological systems

A high proportion of firms in the survey either rate their competitiveness and efficiency poorly or are unable to make comparisons.

Firms showed more confidence in comparing themselves with other local firms with a gradual scaling down of their relative performance or an inability to make comparisons, the wider the network of firms included. This almost certainly suggests imperfections in the information received by firms. Firms, therefore, clearly operate in relative darkness in respect of what constitutes current 'best practice' in the industry and are not in the main at the forefront of technology or production practice.

Just under half the Leicester firms surveyed use some form of computerized office administration system such as word processing, computerized wages, invoicing, payments, stock and/or production control systems. Less than a third of the firms use any form of computerized manufacturing system and only a small minority have a computer aided design system. Firms who are not computerized in any area have not, in significant numbers, even considered using a computerized system. The pattern of use suggests that a computerized administrative system is the most likely to be adopted and could normally be expected to precede any other form of computerization (though one firm has a CAD system and does not use any form of computerized office system).

Our analysis of the survey findings suggests that the major factors influencing the adoption of new technological systems are the size of firms, with larger firms being much more likely to adopt than small firms (100 employees being a critical size); and whether or not firms supply chain stores, those firms so doing being much more likely to have adopted than otherwise. This would suggest that the chain stores exert a certain amount of pressure on their supplying firms to conform to quality and delivery times which puts pressure on firms to operate using the most up-to-date technology. Knitwear firms were found to be more likely to have adopted new technological systems than other firms, reflecting the advances which have developed in knitwear design and production machinery as well as the demands of quality, delivery and colour matching requirements in this sector.

Case study interviewees suggested that the main benefits of computerization of production were the opening up of new markets, improvement in quality of the finished product and the speed and flexibility offered by the new machinery. They suggested that the main disadvantage of using computerized systems was the resultant overdependence on outside expertise. Those companies who had adopted did however feel that the systems had been worthwhile and had lived up to their expectations, though many had found hidden costs such as the inability of the equipment to cope with low quality materials. In terms of office administration systems, firms reported problems with manufacturers' back-up services and overenthusiastic claims by sales representatives of the suitability of hardware and software for the firms' specific purposes.

Introduction of a shared facility of computerized systems

Around half of the firms in the Leicester textiles and clothing survey would consider using a shared facility which offered the use/ demonstration/advice of computerized systems and a similar proportion would consider using a training facility. This is by no means an overwhelming response, given the nature of the offer. Such facilities have, however, been introduced elsewhere (e.g. Birmingham) where they have proved successful and their use in engineering industries has been widespread, again with much success. In view of case study comments in respect of computerized administrative systems and also because it has been shown that office administration is usually computerized before design and manufacture, it would seem logical to consider facilities in this area as a priority over design and manufacture.

The case study interviewees suggested the following features, in order of priority, of any shared facility which might be introduced in Leicester: 'hands on' experience of computerized equipment; information and advice on available technology which was company specific, neutral, expert, and backed by experience of the industry; similar advice relating to technology appropriate to the firm's needs; and finally, a short term rental service for use on company premises.

Although we did not ask any questions relating to the potential for local authority involvement (other than relating to a shared technology and training facility) in the survey of firms, we pursued this in the Leicester case study interviews. Firms on the whole were reluctant to make suggestions for local authority intervention and several felt it inappropriate that the authority should intervene. Clearly, there may be an underlying fear both with the tax rate implications of such intervention but also with the power and autonomy of the small firm. However, in addition to the suggestions made in relation to a shared technology facility referred to above, the following suggestions were made: (i) promotion of the industry to improve its image; and (ii) promotion abroad of the industry and the city. These two suggestions gained a majority of support from the case study companies. Such measures would not only help to increase the demand for local products, but would also help in the recruitment and maintenance of a skilled and talented work force. Exactly half of the companies suggested a reduction in the rates.

The nationwide telephone survey and case study interviews revealed a large number and wide range of shared, high technology facilities in the UK. In practice, the West Midlands proved to be a particularly fertile area for investigation, in part because of the growth of such prospects in the area which reflected an attempt to combat the heavy losses in traditional, male dominated employment opportunities. (Nevertheless, it is worth adding that this locality was selected before the Leicester-based project described above was proposed). All of the shared facilities offered a wide range of training and consultative services including 'hands on' experience of high technology systems.

The case study projects were all planned and established within the last six or seven years. The telephone interviews revealed that these were, in certain respects, the forerunners of a larger number of projects due to come on stream in the near future. The planning prior to the establishment of shared facilities is becoming more organized and coherent over time. The stimulus for the projects was often a local unemployment problem. The emphasis of the projects was generally two-fold: first, to increase competitiveness of firms; second, to upgrade worker skills and improve employment prospects.

Evidence about both initial and current funding suggests a somewhat incohesive and haphazard structure and nature of project facility financing. Industry itself appears reluctant to be involved in a pump-priming exercise. The projects take some time to become self supporting financially. The need to become self sufficient appears to lead to some uncertainty or ambiguity, conflicting with the broader perceived 'service' goals of the projects. Project funding, particularly for expensive capital items, was problematic.

Successful operation of the units requires experienced and competent staff. The shared facilities depended heavily on the enthusiasm and drive of the project leaders. While the projects appeared to be working close to capacity, they were generally able to accommodate new demands, although all had plans to expand. The units were very aware of the need to adapt their activities to the requirements of potential users.

The projects appear to be becoming more formally managed and more accountable to some management or steering committee. Strong links were being forged with outside institutions such as employer federations, trade unions, etc., although one possible weak area appeared to be caused by the attitudes of local chambers of commerce.

The clients included firms of all sizes and small businesses were generally well represented among the users. The shared facilities generally made special efforts to attract small firms and there is evidence to suggest that small businesses represent a substantial proportion of the clientele of shared facilities. There are, however, some remaining questions. The small firms were often not that small (often with between 50 and 200 employees). In addition, it is possible that the user firms were often not those in most need but those most able to make use of the facilities. There was also, with some exceptions, a lack of knowledge or experience of minority groups such as women or ethnic minority owner managers.

The project leaders' own perceptions were that they had helped to achieve a successful transition to high technology among their clients, although they saw their role as being broader than this and did not see high technology conversion as a universal 'cure all'. On the other hand, there was a general view that the new technologies have applications in all sectors and constitute an important key to future efficiency and competitiveness. In particular, they saw shared facilities as a means of achieving this transition, particularly among small firms, because of the complexity and multiplicity of hardware and software on the market and the high costs of technology transition.

The survey of user firms confirmed the picture built up from the interviews of project leaders. All of the firms had achieved the transition to high technology, introducing CAD and/or CAM systems. The randomly selected firms tended to be large, only one employing fewer than 200 people, and two were a part of a group of companies. The survey responses indicated strong evidence of user satisfaction with the project facilities and attitude of the staff. The user firms reported significant improvements in their performance in a variety of areas.

On balance, the potential contribution of such shared facilities therefore appears to be considerably greater than the Leicester textiles firms seem to believe. They may prove to be an important mechanism by which small firms can shift from the low skills/low technology to the high skills/high technology route and, thereby, open up access to a more sustainable growth path.

1 Introduction

Small firms, technology and growth

This book focuses on small firms and their potential for growth. The typical small firm that we have in mind has fewer than fifty employees. It is likely to have limited market power, if any, and what it has appears likely to arise from a low level of product differentiation. In general, therefore, we assume that the small firm operates in a comparatively competitive subsector of the market. Of the alternatives, monopolistic competition, with its associated product differentiation and free entry, appears to best approximate the market structure in which the firm operates.

There is clear evidence that there has been a major growth in both self employment and the number of small firms in the UK in recent years (Wilson and Bosworth, 1987). Despite the growth in the size of the small firm sector, considerable concern has been expressed about the ability of small firms to adopt more advanced technology, 'larger firm' forms of organization, and to achieve sustained growth. While the general links between firm size, technology and organization have been explored elsewhere (Bosworth, 1987; Bosworth and Jacobs, 1987), this paper presents an attempt to provide a more formal framework for the analysis of the growth of small firms.

We attempt to illustrate below that there are essentially two ways that a small firm can adapt to its environment. In essence, these can be classified as high technology/high skill and low technology/low skill routes. In the short run both of these routes can be viable. However, it seems unlikely that firms which have evolved via the low technology/low skill route will generate internal pressures for growth. Neither do they seem suited to the adoption and assimilation of more advanced forms of technology or 'large firm' forms of management and organization. As a consequence, unless some outside stimulus is given, such firms are more likely to remain small.

Competition and competitiveness

The opportunity arose to undertake a detailed examination of the nature of competition and the degree of competitiveness among textiles and clothing firms in Leicester. The aims of this project as a whole were somewhat broader, and are reported elsewhere (Bosworth, Jacobs and Lewis, 1988). The survey and case study material reported below examine: the structure and performance of the clothing and textile sector in Leicester, particularly with regard to the nature and degree of competition involved; the production processes in the industry locally; the extent to which new technologies are used in the sector and the main factors which influence their adoption; production measures which improve company competitiveness, market share and response to consumer trends; links between employment, pay, skill and working conditions; the use of CAD/CAM systems and their value to local firms; and the viability of collective provision on a bureau basis.

Shared facilities as a route to adoption

The evidence from the Leicester-based research gives some indication of the low level of technology in many small firms and the barriers such firms face in shifting toward the high technology/high skill route. Shared facilities are one potential source of stimulation. One of the aims of the case study and survey work was, therefore, to investigate the perceived benefits among potential users in the textiles and clothing industry.

The results obtained from Leicester-based firms are complemented by somewhat broader and more general information about shared facilities, collected during the course of an ESRC project during the pilot phase of the New Technologies and the Firm Initiative. The rationale for the latter was that, while a considerable amount was known about various (often government and local authority) initiatives aimed at establishing new small business enterprises, much less was known about programmes aimed at stimulating the technological dynamism and growth of the firm. The aim of the project was to investigate the nature and role of shared facilities in the small firm which might take a variety of forms. Its primary feature, however, would be that it would act as a medium for the transfer of technology, know-how and/or skills to small firms. As we show below, there are a wide range of shared facilities already in existence and a large number of new initiatives which are now springing up, many of them associated with the diffusion of new technologies such as computer aided manufacture and computer aided design (CAD/CAM), across small firms.

Contents of the report

The remainder of the report is divided as follows: in Chapter 2 we consider the relevance of new technologies and associated skills development to the competitive position of firms. There is much evidence to suggest that a failure to keep pace with technological developments and to invest in appropriate training have contributed to a relative decline in the competitive position of many of Britain's manufacturing industries. Chapter 3 looks at some aspects of the debate concerning the growth potential of small relative to large firms. The evidence suggests that there are often major hurdles to overcome before a firm can cross the divide from small to large scale. Chapters 4, 5 and 6 examine the interrelationships between technological development, small firm dynamics, and skill acquisition within the clothing and textile industry. The findings arise from a study of clothing and textile firms in Leicester. Chapter 4 presents the background context of the industry. Chapter 5 contains the results of a survey of firms, and Chapter 6 reports on the findings of some in-depth studies of a small subset of the sample firms. In Chapter 7 we report on a study of shared technology and training facilities in the West Midlands. The five high technology shared facilities studied offer a wide range of consultative and training services, including 'hands on' experience of a variety of systems. Four of the facilities relate to engineering firms and one to the textile and clothing sector. All are relatively new. We discuss their development, financing, operations, user groups, and development plans. We also report on a survey of firms who have used the facilities and discuss their experiences. Finally, in Chapter 8, some conclusions are drawn and some policy implications are considered.

2 Technology, skills and the competitive position of firms

The competitive position

Although a part of the decrease in UK manufacturing actively took place during a period of general decline in manufacturing in other major industrialized countries, a second aspect, and one with potentially more serious long term consequencies, has been the <u>relative</u> decline of British manufacturing industry vis a vis its European, American and Japanese competitors. Since the Midlands is at the heart of British manufacturing industry the long term prospects of the area depend on a reversal of this trend.

A common theme of the sector studies (concentrated in the metal based manufacturing industries in the West Midlands), undertaken as part of the former West Midlands County Council's economic development strategy (see WERU reports 1983 to 1985), has been that the failure of the sectors to keep abreast of the current best-practice techniques has contributed to their relatively poor performance. The poor response to new technological developments can be traced to a variety, often a combination, of reasons such as financial constraints, risk aversion, a lack of familiarity with available techniques, and a lack of the necessary manpower to operate new technology, etc.

The long term prosperity of the area is therefore dependent on technological awareness and the development of an appropriately skilled supply of labour alongside incentive schemes for capital renewal. Consequently the training which is provided needs to be capable not only of responding to the needs of the individual but also to the requirements of industry. Those needs will have to be identified in advance in order to ensure an appropriate period in which to train people.

Recent research by the Policy Studies Institute (Northcott and Walling, 1988) revealed that 63 per cent of manufacturing establishments in Britain, representing 84 per cent of manufacturing employment, are now using micro-electronics in either their products or in their production processes. The West Midlands region is lagging behind in this respect with only 56 per cent of establishments, representing 79 per cent of employment, having adopted micro-electronic based products or production technology. The East Midlands position is identical to that of Britain. These findings are in line with earlier surveys (see Northcott and Rogers, 1984) which show the East Midlands keeping pace with the average rate of adoption, with the West Midlands lagging slightly behind.

Adoption of technological systems

Although present consensus inclines towards an explanation of the relatively poor performance of British industry in terms of long standing cultural inadequacies, somewhat shorter term analysis focuses more specifically on the role of advanced technology in ensuring success and, conversely, on the failure of British firms to keep abreast of best-practice technology. Some commentators (see Carter, 1981) go further and suggest that it is characteristic of British industry that it has failed to target 'winning' products or that it engages in 'a perverse pattern in specialization' which results in increasing reliance on production areas with a low skill content. Britain's problem is seen to be exacerbated by the rapid pace of technological change being set by leading nations such as Japan, the USA and West Germany.

While there would not necessarily be total agreement over this general diagnosis of the state of British industry, it is likely that agreement would be found on the additional problems engendered by the apparent speeding up of the rate of technological change. Nor would there be widespread disagreement on the opinion that in order to thrive and prosper in highly competitive export markets it is essential that domestic firms make appropriate use of the latest available technology (see NEDO, 1983b).

In the last decade some of the most significant advances have occurred in areas susceptible to the introduction of micro-electronically based technologies. This is particularly true of the engineering industry where modern production technologies encompass numerical control (NC), computer numerical control (CNC), and direct numerical control (DNC) machine tools; robots; flexible manufacturing systems (FMS); computer aided design (CAD) and computer aided manufacturing (CAM) as well as integrated CAD/CAM systems; computer control systems and automated storage. However, it is also true of the textiles and clothing industry which forms the main focus of the empirical evidence presented below. Despite the diffusion which has gone on to date, uptake of many of these technologies is still at a relatively early stage as already noted.

Changes in physical equipment also bring with them corresponding changes in the skills required both to plan their use and to operate them. It is clear that the investments in physical and human capital, particularly in relation to skills associated with new technologies, cannot be evaluated independently. Investment in human capital skills is in many cases a prerequisite for investment in physical capital; as such, this investment includes a much higher risk of non-recoupment of anticipated benefits as well as much greater difficulty in evaluating the potential benefits (Lewis and Armstrong, 1986). Links between new technology and skills have been highlighted in many industries. For example, the turnaround in the steel industry since 1980 has been associated with a seven-fold increase in training for staff (Williams, 1988, p. D.14).

The skills required to operate new technological systems based on a whole new concept of operation (e.g. electronic rather than mechanical) are by definition different from those established through the traditional craft training schemes. Yet many of the old skills will still be required alongside new and very different skills. Not only does this affect the formal training process, but it also influences the way in which skills are transmitted through the work force. A whole new transmission mechanism may need to be developed although in many cases a new mechanism will evolve and, through trial and error, the errors may sometimes prove very costly. One of the most constantly recurring themes in the literature on skill require-ments which has been confirmed by our own research, is the need to combine skills to accommodate, for example, the areas of electronic and mechanical engineering. Associated with this is the necessity for flexibility of the work force.

The particular response which is required is not totally pre-determined by the technological hardware. For example, one area of concern has been the labour displacing aspects of new technology. Here, management strategy can play a large part in determining whether de-skilling takes place (if new technology is introduced in such a way as to fragment work) or alternatively, if more sophisticated technology can be used to remove repetitive tasks. Employees will then be able to oversee and control more stages of production and there will be opportunities to enhance skills (Williams 1984, Lewis and Armstrong 1986, Bosworth and Jacobs, 1987).

The relationship between technological hardware and skills is thus a complex one which affects all levels of the skill hierarchy. Successful integration of advanced technology into established production processes depends heavily on the ability of management to make technically informed decisions as well as making the necessary organizational changes and motivating the work force. Management needs to understand the capabilities of new systems and to identify staffing levels and hence skill implications. The technical specifi-cations decided upon by management will in turn determine the way work roles are reorganized (Burgess, 1985). As we indicated above, often the potential for either de-skilling or skill enhancement is present at this juncture but, as Littler and Salaman (1984) point out, if human contributions are excluded from the work systems, boredom and alienation set in. However, there is no doubt that tasks will change; some operator work will continue to be displaced or de-skilled while other work becomes more complex.

New technological systems

Engineering

In engineering, technological developments and their related skill implications have been widespread. The potential of computers to aid the design process is now recognized. They can be used to prepare 'families' of similar, standard or potentially standard drawings. Then, when a new drawing is required, representations of many of its component parts can be recalled from the data memory and used or modified as required. A number of new skills have to be learned by draughtsmen and women since the equipment requires extensive preparation before it can be used to build up drawings. At the same time the demand for more routine draughting skills has lessened.

Production engineers have borne a major responsibility for the introduction of NC and CNC machine tools, their programming and commissioning. The trend has been for hardware costs to come down and for software packages to be developed. The movement to integrated CAD/CAM systems has blurred the distinction between design and production. In addition to drawings, CAD can produce parts lists, schedules and inspection and test data. The CAM link makes use of information from the CAD data base in the preparation of programmes for direct transmission to CNC machine tools. The skill implications of these new systems are considerable. Engineers and technicians from conventional electro-mechanical backgrounds may have to acquire competence in writing or revising software in systems analysis and electronic diagnosis.

Textiles

With the relatively recent advent of the computerized machine, the textile and clothing industry has faced perhaps its biggest 'revolution' throughout its long history. During the past decade, the development of a range of computerized, high technology machinery has brought both advantages and problems to the industry and, with all that has been achieved so far, the search for further progress in this area continues apace. At a recent international trade exhibition held in October 1987 in Paris, 1,270 equipment manufacturers and designers exhibited their latest technology in an exhibition arena covering 146,000 square metres of floor space (Harrison, 1987). The manufacture of high technology textile, clothing, knitting and hosiery machinery has taken on a global identity in recent years and, while Europe still tends to lead in this field of development, many other countries are now involved. Companies in China, Israel, Iran, Portugal, Turkey and Taiwan are now manufacturing modern machinery for the industry, as well as the 'giant' concerns located in West Germany, Italy, France, the USA, Japan, Brazil and Britain (Knitting International, 1987).

The remarkable advances in computerized technology cover a wide range of functions which offer the textile and clothing manufacturer two prime facilities: speed and flexibility. Knitting machines are now able to reach speeds of 2,600 picks a minute, with design patterns printed on up to six magnetic tapes at a time (Swan, 1987). Electronic needle selection enables faster knitting and, importantly, quick design change and new design implementation. The development of compound needles, rotating cam boxes and needle beds, and increased numbers of yarn feeders on both circular and flatbed machines, now up to 148 on some machines, has further increased capacity, quality and speed of knitting. The problem is no longer one of increasing speed, but of controlling yarn tension at these greatly increased speeds (Wheatley, 1987).

For the clothing industry there is, at present, no completely computerized making up technology yet available. Computerized cutting tables are now available where up to 150 or so layers of fabric can be laid out and covered by a polythene sheet which sucks out the air and compresses the fabric into a hard block. A computerized arm moves across the pattern and cuts the compressed layers of fabrics into the requisite shape whilst maintaining exact size specification. Computerized lay-planning and grading machines, which allow highly

specific size grading and prevent unnecessary wastage of cloth, have
made a further contribution to the clothing manufacturing sector.
Electronic sewing machines now run faster and some processes, for
example, trimming and cleaning up, shirtcuff attachment, and some
embroidery work, can now be programmed and run automatically.
However, it is true to say that the bringing together of sleeves, body
and collar still cannot be done by one computerized machine, although
this is likely to be possible in the future. The problems of fully
automated making up relate principally to the wide range of potential
fabrics, all of which behave in a different way, and also to the
porous nature of fabric itself, thus making it a difficult substance
for machines to handle. The fact that garments have to be made to
wide ranges of different sizes also poses problems for computerized
making up processes.

High technology developments are now available at the dying and
finishing stages of fabric processing. Computerized temperature
control mechanisms can now confront the problems associated with
accurate colour matching of yarns and fabrics. Automatic combing and
brushing machines have been developed to give improved quality output
at greater speeds. Fully automated systems are now in operation which
sort garments by size, colour and style, and convey them in batches
into a specially adapted delivery truck for dispatch to the retailer,
ready hung, in near perfect condition, for transfer on to shop display
rails. Such systems are greatly speeding up the delivery of garments
from the manufacturer to the retail outlet, to the extent that goods
ordered on a Monday morning can now be in the store for sale by the
following weekend (Apparel International, 1987). Automated guided
vehicle systems, based on light sensors, can now supply garment pieces
to factory floor work stations, where ticket control systems monitor
the work flow at each of the various stages of the production
process. Future advances for the industry are now concentrating on
robotic development, particularly fabric handling for the clothing
sector of the industry, and robotic yarn handling for the spinning
industry (Parkinson, 1987).

There have, therefore, been significant developments in the design
and production of automated, computerized machinery for all sectors of
the textile and clothing industry and more is undoubtedly on the
way. However, the process of adaptation to this new technology is not
without its pitfalls. In the view of Teague of Software Systems UK,
computerization can be vastly expensive, and basically it is only the
larger companies which are able to take full advantage of such
developments. Adaptation to high technology is only feasible in terms
of cost to firms with around sixty employees or more. For the smaller
firm it is still more cost effective to buy in cheap labour than it is
to invest in sophisticated equipment. Cheap Third World labour is an
exploitable commodity much utilized by many companies in the UK
industry and thought, by some, to be to the detriment of the home
trade (Teague, 1987). Additionally, companies that do invest in the
new technology do not always realize the need to readjust their whole
manufacturing processing system so as to back up the new, rapid
manufacturing facility. With machines which will sew or knit at 6,000
stitches a minute, backlogs can quickly build up and supplies can
rapidly run short. Managements are having to adapt to new systems
which will encompass every aspect of every manufacturing process in
order to achieve maximum capacity from high technology investments
(Teague, 1987).

Although many firms may find it difficult to adopt the new technology, basically because of the cost involved, there is high and growing investment, especially by the knitting industry. In order to compete on equal terms with other firms in the home trade and to produce a wider variety of high quality designs with rapid design change, knitting companies have, in general, invested heavily in computerized knitting machines (Boggon, 1987). The results of our analysis of the survey data, outlined below, support this view that knitting firms are more likely to have adopted computerized machinery than other firms in the clothing and textile sector in Leicester. The non-adoption of high technology for all sectors of the industry may result in company liquidation, or at least restrictions on producing for the lower, mass production end of the market (Loman, 1987). It would seem that these days only the very specialist firm can afford to completely side-step the 'computer revolution'. The challenge to the high technology equipment producers and designers is now to facilitate fully-integrated systems for design, manufacture and work flow management control which can enable the industry to respond faster and more effectively to market demand without incurring expensive production cost mistakes (Millington, 1987; Teague, 1987).

Skills and training

Textiles and clothing

Although the production process remains labour intensive and the making up of goods still involves the use of a stitching machine operated by a machinist in the vast majority of cases, the clothing industry today includes many innovative enterprises. The Clothing and Allied Products Industry Training Board (CAPITB) reports a number of companies with recruitment difficulties. Although the IER forecast is for a slight downturn in employment, there may well be difficulties in recruitment of young people with particular qualities and skills, especially in the area of design (IER, 1990) In textile production too, the use of more technically advanced, micro-electronic based production techniques may cause problems in recruitment of individuals with appropriate skills (such as programming skills). Such people are in heavy demand in other sectors of manufacturing industries which have higher rates of pay.
There is perceived to be an urgent need for the industry to improve its image, partly by putting its own house in order as this relates to the working conditions of its staff and the quality of its management functions, and partly by an increased investment of time, effort and money into self advertisement, particularly in respect of contacts with the export market. As the export market now assumes an increasing importance for the industry, it becomes commensurately more important to create and stabilize a reservoir of skilled, committed people. Designers especially are expected to fulfil a crucial role in the success of both the home and export trade for the UK industry in the future. It would seem entirely counterproductive if an industry which in looking to the future, has made determined efforts in terms of investment and adaptation to changing circumstances, should continue to be blighted by an image which can only limit its future potential success.

Engineering

In engineering the EITB has indicated that such developments have led to a large increase in demand for multiple-skills experience. It is now common for technicians to work as members of design and development teams and the stress is on combining theoretical knowledge and practical skills. Consequent upon these changes, there is evidence of polarization of skill levels. The more routine elements of traditional technicians' work have been automated, leaving individuals either to move up the employment hierarchy to undertake more complex, non-routine work or, conversely, to find that their skills have become redundant. This raises the important issue of how retraining or upgrading of skills may be assisted since the conventional training routes for technicians may no longer be appropriate.

The impact of new technology on production workers has been largely determined by the drive for lower unit costs. This in turn has led to the search for reducing the time spent on the production process, for example in the metal working industries in fitting, forming and machining, finishing and rectification. More specialized machines mean greater accuracy, higher quality and throughput. Metal cutting, forming and bending machine tools have progressed from manual controls through CNC to machining centres and integrated materials handling systems. In some cases machine instructions are produced entirely by technicians with a consequent reduction in the use of shop floor craft skills.

The net result may be that time served craftsmen and women end up supervising CNC machines. A study undertaken at Imperial College, London (Francis et al., 1982) refers to the strict demarcation between craft workers, machine operators and part programmers. In some cases, craft workers learnt the necessary programming skills and unofficially made adjustments to programmes. Company reaction to this was mixed. Some firms encouraged the development of programming skills to enable them to exercise more control over the machine tools. Others locked up the programming controls to prevent them from doing their own programming. Thus, in this area, there is also an increasing polarization between mundane machine minding, and skilled systems control and fault diagnosis.

Changes in the area of maintenance skills have been particularly marked. On the one hand, new technological systems may be inherently more reliable; on the other, the complexity and intensity of use may place added demands on maintenance skills. Again, a combination of theoretical knowledge and practical skill experience is regarded as the desirable combination, although there is less certainty over how it is to be achieved. The lack of high level maintenance skills for micro-electronic based production technology is one of the most frequently recurring themes in the literature we have surveyed (e.g. Senker, et al., 1981).

Finally, manual skills are those most often identified as being particularly vulnerable to displacement by automated processes. However, some reconsideration is occurring over the nature of 'strategic' skills - strategic in the sense that the degree of precision to which they are developed and the nature of the products on which they are used, may be almost unique to a particular industry. Such skills (machining and fabrication skills are suggested as examples, see NEDO 1983b) may need to be maintained and practised even if automation appears to be capable of eliminating their day-to-

day use since, once lost, they cannot quickly be regained. The full
implications of such losses are as yet unclear.

Small Firms and Shared Facilities

The developments and adoption of new technological systems within
industries such as engineering will lead to a widening gap between
adopters and non-adopters. Inevitably the smaller firms are those
most likely to be among the non-adopters; they generally have less
finance available for investment, often borrowing at higher rates;
they usually have less expertise and knowledge at their disposal; they
frequently have less time to make available to investigate alternative
systems; and the risks to investment in both physical and human
capital are relatively greater (Metcalf, 1987).

In the textile and clothing industry these experiences are likely to
be repeated. In this industry the problems of adoption of new
technologies may be greater due to the larger number of small
production units. As we have already noted, smaller enterprises
experience greater difficulties and costs in adopting new
techniques. In this respect the gap between (usually large) adopters
and (usually small) non-adopters is likely to widen. Furthermore, the
difficulties associated with transition from small- to large-scale
production are likely to increase.

As we have already noted above, some investment in gaining
information in relation to new technologies at management level is a
necessary prerequisite to the introduction of any new system. The
existence of a Resource Centre or other similar shared facility is, in
principle, an extremely valuable method of facilitating such changes
and reducing the time input into this process. The problem of the
lack of expertise within many firms, and especially within smaller
firms, can be offset by the availability of expertise within the
shared facility. Shared facilities also enable the risks of
investment in both human and physical capital to be substantially
reduced and more easily evaluated.

3 Dynamics of small firm behaviour

Barriers to growth in small firms

Concern has recently been expressed about the existence of barriers to growth among small, potentially innovatory firms. This has been the focus of attention in a recent ACARD study (Metcalf, 1987) which isolated a number of potential barriers. The first of these concerned managerial attitudes, behaviour and abilities. The second concerns potential problems of access to, or cost of, key resources such as finance, skilled labour and technology. Finally, there are questions concerning whether current market structures, including large firm/ small firm and government/small firm relations, put small firms at a particular disadvantage. The essential problem areas which appear important in the context of this study concern the availability and relative costs of resources to small and large firms.

There is substantial evidence, for example, that the bulk (perhaps as much as 75 per cent) of capital finance for the small firm comes from internally generated funds (Wilson, 1975; Binks and Vale, 1984; Hall, 1987). Where these funds are inadequate, the small company tends to turn to the Clearing Banks (perhaps 12 per cent of finance is raised in this way) and small firms face around a 2 per cent higher rate than large firms (Wilson, 1975, Hall, 1987). The ability to generate sufficient funds to finance growth is incorporated in certain strands of the managerial theory of the firm (Marris, 1963, 1964). Our concern is that the capital intensity of certain technologies may put them out of reach of many small enterprises, thus providing an effective barrier to growth. The costs of hardware or software often make them uneconomic to buy unless the firm is sufficiently large to ensure that they are utilized intensively.

While small firms face lower per unit labour costs, this appears to be at the cost of a lower quality of work force (Bosworth, 1987). It has been argued that the quality of the work force may be a major determinant of the ability to adapt and adopt new technologies (Solo,

1966; Johansson and Karlsson, 1985; Bosworth, 1987). Thus, the adoption of a low quality work force option may seriously impair the small, potentially innovatory firm's ability to innovate and grow. This is further reinforced by the lower level of training per worker offered in smaller firms (Bosworth, 1987). A related influence on performance is the quality of the management measured in terms of its attitudes, behaviour and abilities (Bosworth and Jacobs, 1987). This may have as much to do with the financial and commercial abilities as with the technical knowledge and skills of the management team. Technical expertise has a high cost, either when hired as an employee or when acquired by the manager through education and training. Thus, access to specialist knowledge almost certainly implies a higher level of expense, initiative and risk for the manager of a small enterprise than a large firm. Even if the owner/manager is highly qualified when he or she sets up the business, there is the continuing problem of updating knowledge and skills.

There are a number of puzzles to be sorted out. Something of a question mark exists over the level of technology adopted by small firms. The general view appears to be that they are technically less efficient than their larger counterparts. Casual observation of investment data (Business Statistics Office) and capital stock data (Metalworking Production) suggests that, on average, they are less capital intensive and have fewer modern technologies (see also the diffusion data contained in Northcott and Rogers, 1984). On the other hand, studies of their comparative efficiencies have not found significant differences (Todd, 1971). There is a widely held view that smaller (often owner controlled) firms tend to be more profitable but grow more slowly than their larger (often professionally managed) counterparts (Samuels and Smyth, 1968, Singh and Whittington, 1968, Todd, 1971, Cable and Steer, 1978). This finding is also confirmed by US studies, particularly when loss making units are excluded from the sample (Todd, 1971, p.6). One suggestion, which probably did as much as anything to spawn the major managerial theories of the firm literature, is that the change in emphasis from profit to growth occurs because of the movement from owner to professional management.

What we are lacking is an analytical framework within which we can piece together the various parts of the small firm jigsaw puzzle. First, why is it that low technology/low skill (often small) firms can coexist alongside high technology/high skill (often large) firms in the same industry? Second, why is it that smaller firms may find it more difficult to grow? Third, how do changes in technology, skills, size and organization mesh together?

Coexistence of high and low technology firms

Figure 3.1 is based on the established production region, G, of the short run production function (Johansen, 1972). The coefficients E_1 and E_2 represent the input of labour per unit of output and capital per unit of output, respectively. We assume for a moment that all of the firms produce an identical product using different technologies. Unlike the original Johansen (1972) work, it is assumed that capital and labour markets are both segmented according to the size of the firm (for example, lending institutions might set guidelines on conditions for loans for different sizes of company). The lines represent the zero profit per unit of output line for each size of firm. It is assumed that large firms are faced by low interest rates

and high per unit labour costs. Small firms are faced by low per unit
labour costs and high interest rates.

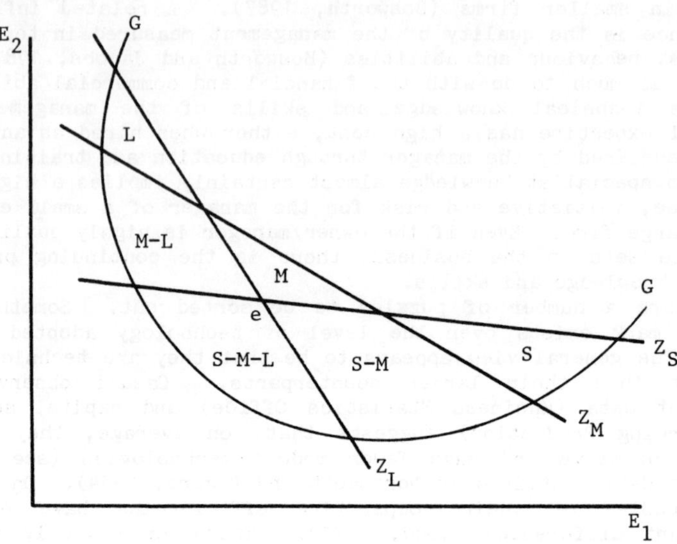

Figure 3.1   Technical efficiency per unit cost and profits of small,
             medium and large sized companies

Figure 3.1 illustrates the coexistence of small, labour intensive
enterprises alongside large, capital intensive firms.   Only small
firms survive in the region S, medium sized firms in M and the largest
firms in L.   Combinations of firm sizes are possible in the other
regions.   It is difficult to isolate the relationship between firm
size and capital intensity without knowledge of the historical nature
of the distribution.   However, if it is assumed that there was
orginally a random distribution of firm sizes across the region G, the
imposition of different factor prices in different segments of the
market would produce an outcome in which size tended to be smallest
for the most labour intensive firms and to increase with the capital
intensity of the technology.   Large firms would not continue to
survive in region S and small firms would disappear from region L.

In practice, it is possible to show that the profit goal will
further shift firms in the various subregions of G. A small firm at
point e, for example, will recognize that per unit profits and total
profits are higher at that capital/labour ratio for medium sized firms
than for small firms (or, for that matter, for large firms).   Thus,
there is an incentive to move to the next size group. This is also
true for some small firms in the S-M region of Figure 3.1.   This
reinforces our earlier result about the relationship between capital
intensity and firm size.   It also helps to show why studies of
efficiency based on the relative costs of small and large firms (Todd,
1971) tend to be inconclusive, because variations in inputs per unit
of output are offset by differences in factor prices.

A model of production and growth

Before extending the analysis to look at the relationship between technology and growth, we turn to the nature of the firm's production function and technical change function. In this section we attempt to illustrate the essential choices faced by small firms and the dynamics of small firm behaviour. The internal stimulus for faster growth comes from a diversion of resources from current production and profit generation, towards dynamic activities such as research and development (R&D), advertizing, etc.

The current production function can be written,

$$Y_t = f(X_t, Q_t, T_t) \tag{1}$$

Where: Y denotes the current output of the firm; X is a vector of factor inputs; Q is an associated vector representing the quality of inputs; T is the proportion of their operating time, U, devoted to current production rather than dynamic activities; t denotes the time period.

By implication, in this simplest of models, the dynamic performance of the firm can be represented by,

$$d_{t+1} = g(X_t, Q_t, V_t) \tag{2}$$

where: d denotes the change in performance between periods t and t+1; V is the proportion of their operating time, U, devoted to dynamic activities rather than current production. Equation (2) therefore represents the 'dynamic' production function. This general dynamic function contains within it the R&D production function.

It has sometimes been argued that an important difference between small and large firms is associated with the greater degree of diversification among larger firms and thereby a greater ability to use the 'unexpected' research outputs of the firm. Given that the extent of diversification is likely to be quite low among small to medium sized firms, this seems unlikely to be a significant factor. In addition, it may even be a suspect argument, given the existence of a strong patent system and the ability to licence. More important in the context of small firms appears to be the fact that larger size allows economies of specialization and scale in the R&D function. Thus, for any given level of dynamic activity per unit of output larger firms are able to separate static and dynamic activities (i.e. for some sub-set of X in equations 1 and 2, T=0. However, R&D is only one aspect of the technological change process, and it would be misleading to separate the static and dynamic activities into two mutually exclusive groups. The successful applications of new technologies within the firm may well depend on the ability of production workers to adopt and assimilate the new technology. This ability will depend crucially on the quality of direct production workers.

Clearly there are similarities with the type of growth model developed by Marris (1963; 1964) which is essentially a model of a large managerial enterprise. In the Marris model, the performance variable, d, is proxied by the rate of growth and this depends on the rate of diversification and the success of diversification. The latter is determined by the retained profits of the firm which are devoted to R&D and advertizing. In equation (2) above, the extent of

dynamic activities is represented by the amount of time, V, devoted to
them by each of the inputs, X. The output of the dynamic activities,
measured by the rate of growth of the firm, depends on both the
quantity and the quality of the inputs. The potential extent of
dynamic activity, as in the Marris model and many of the structure-
conduct-performance models, depends on the retained profits of the
firm (Mueller, 1967; Grabowski and Mueller, 1978).

Criticisms of the Marris model include the assumption of growth
maximization (particularly for the suppliers of capital), the
concentration on diversification (rather than growth within a given
market) and the failure to account for competitive dynamic activities
of other firms (see Koutsoyiannis, 1975, pp. 352-70, Dasgupta and
Stiglitz, 1980a, 1980b). In the context of small firms, growth within
a given market appears potentially more important than diversification
behaviour. The failure to account for the R&D strategies of
competitors, which is the principal theme of Dasgupta and Stiglitz
(1980b), appears less important in the context of small firms. With a
large number of competing small firms, competitive R&D activity can be
approximately represented by measuring X, Q and V relative to the
industry averages, which are independent of the individual firm's
decision.

Profit maximization and growth

For the small firm the question of growth maximization appears
subservient to the maximization of the discounted sum of future net
profits, although the latter clearly has implications for the optimal
growth rate of the enterprise. In practice, for the monopolistically
competitive firm restricted to growth without diversification, the two
will be closely connected. Reduction in cost with a given demand
curve leads to increased output (i.e. positive growth) and higher
profits. Increases in demand at given cost levels produce higher
output (i.e. positive growth) and greater profits. Clearly, however,
if the firm attempts to grow at a more rapid rate than that determined
by the downward movement in costs or the outward shift in demand, then
faster growth results in lower profits.

Figures 3.2 and 3.3 show the main alternatives open to the profit
maximizing firm. Figure 3.2 shows the case of a process change, with
given demand D. The effect of the improvement is to reduce costs from
c_0 to c_1, increasing profits from π_0 to π_1, and causing the firm to
grow from Y_0 to Y_1. Thus, the rate of growth in each period depends
essentially on the rate of cost reduction (as well as the slope of the
firm's demand curve and the immediacy of its competitors'
responses). Figure 3.3 shows that a similar result applies in the
case of a product change, with given costs, c. The improvement shifts
the demand curve for the product from D_0 to D_1, raises profits from $?_0$
to $?_1$, and gives rise to a growth in output from Y_0 to Y_1. In this
case, the rate of growth in output is determined by the rate of
product improvement.

The question arises whether it pays the small firm, having made an
invention, to take over an increasing part or even the whole of the
industry? The answer to this turns on whether the invention is
drastic or non-drastic. For non-drastic inventions, given the
presence of a patent system which ensures appropriability of the know-
how, the returns to the inventing firm are greater from licensing
to all firms in the industry, than to transforming the industry from

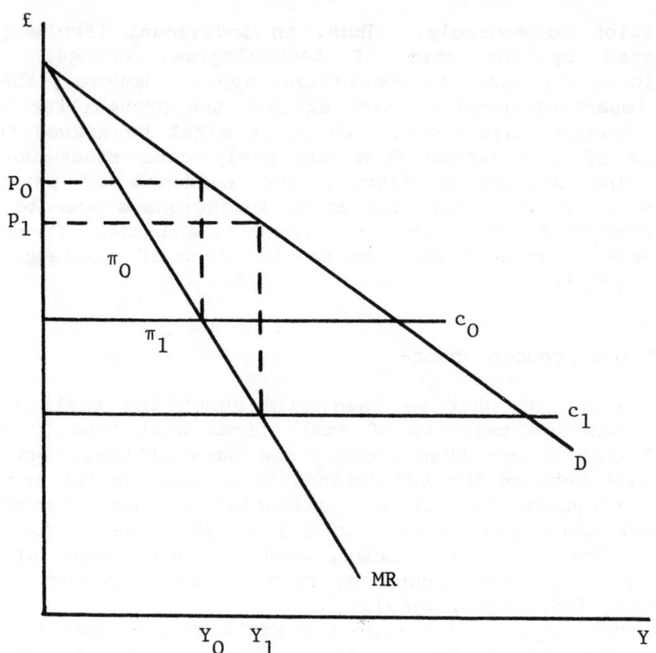

Figure 3.2 Process change and growth

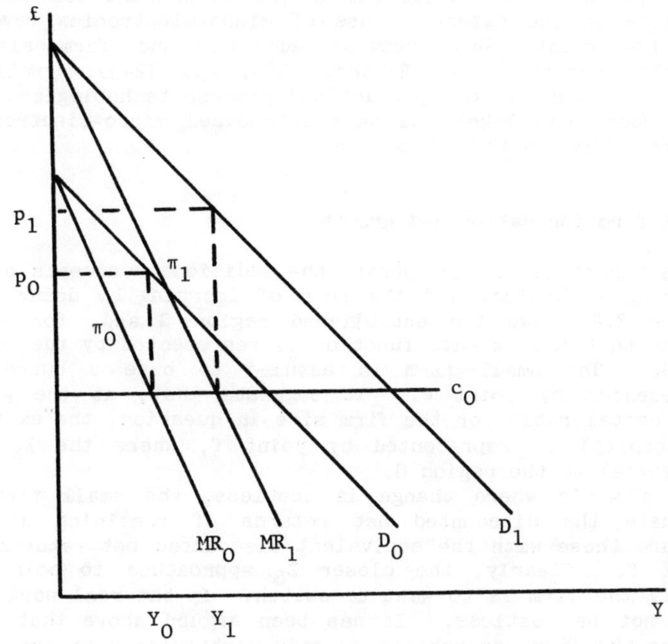

Figure 3.3 Product change and growth

competition to monopoly. Thus, an individual firm's growth remains restricted by the rate of technological change. For drastic inventions, the opposite conclusions apply. However, there appears to be an important question mark against the probability of small firms making drastic inventions. While it might be argued that there are examples of the latter from the early case study work of Jewkes, Sawers and Stillerman (1969), the evidence of patent statistics appears to indicate that the major inventions appear to come from the large corporations; individual inventors and small firms tend to make relatively minor contributions to the stock of knowledge (Bosworth and Wilson, 1987).

Product and process change

In the light of what we have said about the small firm, it seems likely that the majority of small firms will tend to adopt existing technologies rather than produce new ones of their own. In essence, the choice between the two depends on a cost benefit evaluation of the invention/innovation and the imitation routes. Nevertheless, the framework developed in Section 3.3 is still valid for the imitation route. The costs of change, however, are those of adopting and assimilating the new technology rather than producing a new technology (Giliches, 1957, 1960, 1980).

A further question concerns the small firm's choice between product and process improvement. Again there is little evidence on the relative importance of investment in product and process change by all firms, let alone by small firms. (The last attempt to collect data about product and process R&D in the UK was the CBI survey of 1958). Evidence on the extent of use of micro-electronics reveals a strong positive relationship between adoption and firm size (see, for example, Northcott and Rogers, 1984, pp. 32-3). While this basic result is true of both product and process technologies, smaller firms were much less likely to have introduced micro-electronics in their products than in their processes.

Small firm innovation and growth

It is important to incorporate the additional concepts of the costs of changing technology and the role of learning by doing in the model. Figure 3.4 shows the established region G and, for simplicity, we assume that the ex ante function is represented by the convex boundary of G. The small firm is assumed to have a current technology represented by point e. It suggests that, at the given (future) wage/rental ratio for the firm size in question, the ex ante choice of new capital is represented by point f, where the Z_S line is just tangential to the region G.

In a world where change is costless, the small firm will simply evaluate the discounted net returns of remaining at point e and compare these with the equivalent discounted net returns of moving to point f. Clearly, the closer Z_S approaches to point e, the more likely the firm is to want to switch. In the real world, this switch will not be costless. It has been argued above that the amount of change the firm can achieve depends particularly on the quality of its work force. For example, there may be a distinction between the amount of change the firm can achieve with a given work force, depending on whether the change sought is in terms of increased efficiency at the existing capital intensity (i.e. along the ray through e) or in terms of changing the capital intensity (i.e. moving to rays closer to the one through f) (see Bosworth, 1976). The larger the change, the greater the change in the knowledge base of the firm and the longer learning by doing can be expected to take. This can be approximated by a quadratic cost of adjustment function as the firm moves from point e along the ray. In addition, there is a further cost of moving away from the ray Oe towards Of. Again, this might be represented by a quadratic cost of adjustment function. Thus, the firm will calculate the discounted sum of marginal costs and benefits along each route and will choose to adjust some way towards f.

Traditional diffusion models have tended to assume only the choice between the existing technology and the most modern. In practice it may be possible to buy intermediate technologies; in fact the firm may be able to resort to buying second-hand intermediate technology. The ability to buy second-hand machines appears particularly important in the context of small firms, and we return to the more general implications of this below. If only the most modern technology is available, the firm with a number of machines and workers can still approach point f gradually by replacing only a part of its existing machine stock in each period. The latter is no more than the intra-firm diffusion process. Access to even a small number of machines within the firm may facilitate improvements in work force quality and the ability to assimilate new machines. In part, this may help to explain the early exponential growth path of the intra-firm diffusion process.

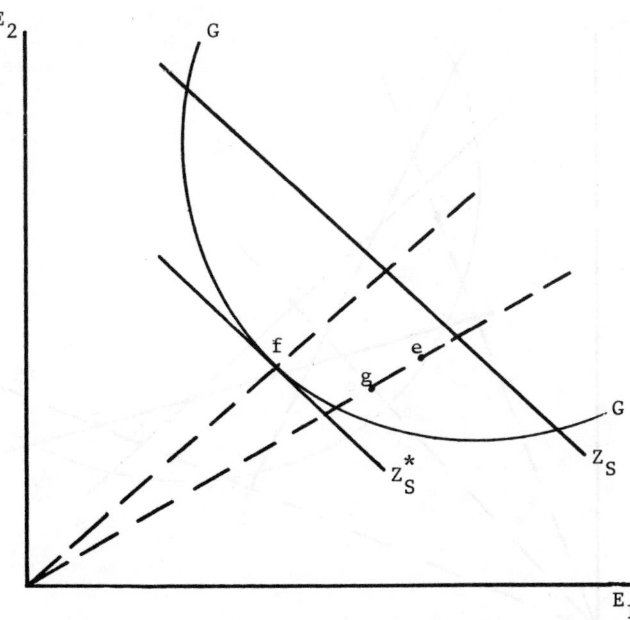

Figure 3.4   Costs and benefits of adjustment towards best-practice techniques

Changes in technology and firm size

Improvements in the firm's technology relative to the industry average causes growth. It is possible to see that the improvements take the firm to technologies which, even if they are not profit maximizing, can support a medium sized firm. See, for example, the movement from point a to point c in Figure 3.5. Changes in firm size also carry costs of adjustment. These are well documented in the factor demand literature (e.g. Hazledine, 1979). In particular, there are costs of adjusting the number of workers and introducing new forms of organization (i.e. large-firm management techniques and organization) and work patterns (e.g. shiftworking).

The potentially most radical change is associated with a switch of both technology and firm size. For example, the small firm might compare the costs and benefits of moving from point a to point c in Figure 3.5 with the costs and benefits of the analogous best-practice point for medium (or large) companies. If the potential discounted stream of future profits is higher at larger company sizes, this provides an incentive to change the capital/labour ratio, move to more advanced technologies (i.e. closer to the frontier function) and increases the size of the company. Clearly, such a move carries with it the 'treble' costs of adjustment (associated with changes in capital intensity, firm size and technological efficiency).

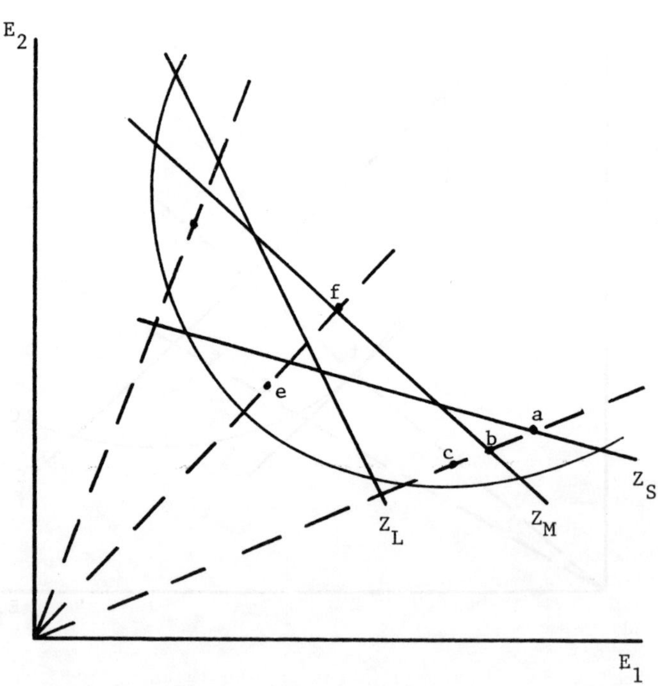

Figure 3.5   Switches in technology and firm size

Technical progress and technical regress

There are clearly a variety of reasons for the recent strong growth of the small firm sector. It is partly a reflection of the general political ethos favouring small firms and enterprise. However, there are two recent developments which relate particularly to the framework outlined in earlier parts of this section. The first of these concerns trends in the casualization of the labour force. The second is associated with the impact of the recession (1980/81) on the scrapping and second-hand sales of machine stocks.

If we re-examine the zero profit line, we find that:

$$p = wE_1 + rE_2 \qquad (3)$$

where: p is the price of the product; w is the user cost of labour; r is the user cost of capital; and E_i are the input/output coefficients for labour and capital. Increased competition in the industry forces down prices, $dp/dt < 0$, and this has important repercussions for marginal firms. In our discussions to date, we would assume that they are either eliminated from the market or are forced to adopt cost-reducing technological change, $dc/dt < 0$ (such that $dc/dt \leqslant dp/dt$). Thus, to date, we have assumed given factor prices and the total change in costs can be written,

$$dc = (\partial c/E_1)\, dE_1 + (\partial c/E_2)\, dE_2$$
$$= w\, dE_1 + r\, dE_2 \qquad (4)$$

Thus, the required improvement in total costs is brought about by a weighted average of the factor-saving changes, where the weights are the factor prices.

An alternative firm strategy, however, is to search for cheaper sources of factor inputs. If for the moment we assume that the firm can apply these inputs on existing technology, then E_i are assumed constant and the total change in cost can be rewritten,

$$dc = (\partial c/w)dw + (\partial c/r)dr$$
$$= E_1\, dw + E_2\, dr \qquad (5)$$

where the required cost reduction is brought about by decreases in w and r. In practice, small firms may find it easier to locate cheaper sources of labour than finance, leading to a further 'twist' in the zero quasi-rent line, inducing them towards even more labour intesive processes.

The two cases of technological improvements and factor price savings appear as alternatives. They are alternatives because the lower priced factor inputs tend to be of lower quality and this increases the costs of making technological improvement and, therefore, make the latter less likely. Thus, the firm examines the costs and benefits of the two routes and selects the one for which the discounted sum of future profits is larger. Under fairly stable technology and factor price conditions we would expect to find a fairly stable proportion choosing each route. However, the recent rapid increases in low priced foreign competition appears to have forced this decision on a larger number of firms, many of which proved ill equipped to handle rapid technological change. A consequence of this intense competition has been the large number of closures leading to sales of capital assets, the evolution of a large pool of unemployed and the casualization of the work force. The effects of this have been to alter the costs and benefits of the two routes and there may well have been a movement away from the high technology/high skills route and towards the low technology/low skills route.

Firms have placed less emphasis on the introduction of the most advanced technologies. They may have kept machines for longer and purchased second-hand machines coming on to the market from other firm closures. While they have faced the consequences of this in terms of higher maintenance costs, this may have been offset to some extent by the lower costs of both parts and labour. The results of this strategy for technological change are not immediately clear. To the extent that the acquisition of second-hand machines is an improvement on the earlier technologies of the firm, the improvement may not be as rapid as it would be if the firm were effectively forced to come to grips with the most advanced technology. In addition, any extension of the user life of capital is likely to have deleterious effects.

These trends appear to have enabled firms to spend less on training and to employ lower quality workers. The result is a downward spiral of technological capability because the average quality of the labour force is not increasing so rapidly. While production can continue for some time, this scenario is worrying because the scope for savings in terms of factor price reductions appears limited and should be viewed against the inexorable downward pressure on product prices arising from the technological improvements adopted by our foreign competitors. As the limits of such savings are reached, future improvements in the form of increased productivity must be achieved from a lower relative technological base. The costs of future adjustment are that much higher.

Shared facilities

If this view of the world is correct, some mechanism is required to reverse this downward spiral which appears to lead to a dead end, where surviving firms are least well equipped to respond by improving their level of technology. An appropriate mechanism appears to be one which reduces the costs of undertaking the transitional step from low to high technology. One possible mechanism for achieving this is the provision of shared facilities which are capable of providing advice and training, as well as 'hands on' experience of potentially appropriate new technologies.

4 Background to the textile and clothing industry

An historical perspective

The manufacture of cotton and woollen yarns, fabrics and garments has developed from primitive peasant art-form origins, and expanded into what is now the fourth largest UK industry, with an annual turnover of £11,700 million in 1985 (Textiles Statistics Bureau, 1986). This development, far from being one of steady growth within a stable environment, has involved the industry in recurrent adaptation to major fluctuations in demand and significant technological innovations.

For 700 years the industry formed one of the principal 'pillars of the State' when wool, the oldest material to have been utilized for textile use, provided the chief manufacturing base of the country. From a nationally dispersed medieval household occupation, woollen cloth making developed as a trade, gradually becoming concentrated in specific geographical regions. One of these areas centred around Leicester, which has been known as a wool town since the twelfth century, and continues as a principal base for the modern knitwear industry today (Lipson, 1953). Much of the textile sector relocated to areas such as Lancashire where there was a source of water power for the power looms, leaving the clothing and knitwear firms to continue to develop in Leicester.

As a comparative newcomer, the cotton industry, more closely associated in the UK with the start of the industrial revolution in the eighteenth century, has an historical base in the cotton growing countries of India, the Near East, the West Indies, Japan and Turkey, with the European industry acquiring its raw cotton mainly from the West Indies and Turkey. In the first half of the eighteenth century UK cotton consumption was estimated to be around 2 million pounds (weight) a year, but from 1760 onwards, consumption increased steadily and peaked around 1913 at about 2,100 million pounds (Robson, 1957).

As with all large industries, the textile and clothing sector of the economy has felt the effects of changing national investment priorities and worldwide trading conditions. The depression years of the 1930s marked an important turning point for what had traditionally been a predominantly home-based woollen industry serving a domestic market. During this period the UK share of the wool export market fell by one-half. For example, the UK share of the Far East market fell from a peak of 18 million pounds (weight) in 1928, to 2.5 million pounds in 1938, due primarily to concentrated investment by the Japanese home-based woollen industry (Lipson, 1953). The post-war history of the knitting industry has been described as being one of feast and famine, with the industry being paid to destroy many of its looms in the 1950s (Parkinson, 1987), recovering somewhat in the 1960s, then experiencing recession again in the 1970s. In the view of one commentator, cheap imports from the Far East, some of them of good quality, exacerbated knitting industry difficulties during the 1970s, and EC, Korean and Indonesian products pose a new threat for the late 1980s (Millington, 1987). Cheap imports from the United States also pose threats to the industry. The prospect of complete harmonization of the EC by 1992 and the possible introduction of VAT on children's clothing are likely to have at least short term effects on sales. The single market in 1992 presents both a challenge and an opportunity for British firms.

The establishment of mills in the cotton producing countries in the early part of this century, built and operated with UK technical expertise and machinery, signalled a continuing decrease of Britain's share of the world market centring on the UK loss of the Indian and Japanese export markets. From the depression years Britain emerged with one-half of her pre-depression output. While all the major Western countries, and the cotton producing countries, invested in their cotton textiles industries, Britain did not, preferring instead to invest in war expansion industries. Britain failed to regain a manufacturing initiative after World War II, due mainly to recruitment difficulties among a disenchanted labour force. By 1950 Britain's share of the world cotton export market, which she had once dominated, was down to 15 per cent (Robson, 1957). Cheap imports from India and the Far East dominated the UK cotton industry during the 1950s, 1960s and 1970s. A new threat has emanated from the poorer EC countries in the 1980s in terms of cheap garments, and from the more developed EC partners in terms of style and quality design.

The experience of the textile and clothing industry since the beginning of the 1970s has been one of consistent cutbacks in large-firm output. Increasing imports and decreasing exports have resulted in a £1.8bn trade gap incurred by the industry during the past few years (Totterdill and Pearce, 1986). Textile and knitting company liquidations have accounted for a disproportionate amount of all UK liquidations in recent years, highlighting the uncertain nature of the industry's environment. This generally depressing situation would seem to be continuing, with import and export figures revealing a widening of the trade deficit to £1,311 million in the first half of 1986 (Textile Horizons, December 1986).

The structure of the industry

Both the textile and clothing industries can be divided into a number of subsectors. The 1980 Standard Industrial Classification (SIC) relating to the industry is contained in the broad (single digit) industry division 4, 'Other manufacturing industries'. Textiles then form the subgroup (two digit) 43, and clothing and footwear form the subgroup 45. The breakdown of the industry is illustrated in Table 4.1 below.

Table 4.1
Structure of the textiles, clothing and footwear industries

SIC no.	Industry breakdown
43	**Textiles**
4310	woollen and worsted;
4321	spinning and doubling of cotton etc.;
4322	weaving of cotton, silk and man-made fibres;
4336	throwing, texturing etc. of continuous filament yarn;
4340	spinning and weaving of flax, hemp and ramie;
4350	jute and polypropylene yarns and fabrics;
4363	hosiery and other weft knitted goods;
4364	warp knitted goods;
4370	textile finishing;
4384	carpets;
4385	other textile floor coverings;
4395	lace;
4396	rope, twine and net;
4398	narrow fabrics;
4399	miscellaneous textiles.
45	**Clothing and footwear**
4510	footwear;
4531	weatherproof outerwear;
4532	men's and boys' tailored outerwear;
4533	women's and girls' tailored outerwear;
4534	work clothing and men's and boys' jeans;
4535	men's and boys' shirts, underwear and nightwear;
4536	women's and girls' light outerwear, lingerie and infants' wear;
4537	hats, caps and millinery;
4538	gloves;
4539	miscellaneous dress industries;
4555	soft furnishings;
4556	canvas goods, sacks and miscellaneous made-up textiles;
4557	household textiles;
4560	fur goods.

Both the textile and clothing sectors are characterized by a disproportionately large number in employment, located within small production units, compared to manufacturing as a whole. The Census of Production indicates that in clothing 13 per cent, and in textiles 14 per cent, of the work force are employed in units of fewer than 10 employees (see Table 4.2 and Figures 4.1, 4.2 and 4.3). This compares with only 10 per cent in manufacturing as a whole being in this size of production unit. In the 10-19 employee band the difference is greater, with 21 per cent of the clothing work force and 24 per cent of the textile work force employed, compared with only 14 per cent for manufacturing as a whole. In the next size band, 20-49 employees, 27 per cent of clothing workers are employed, 29 per cent of textile workers and only 20 per cent of manufacturing workers. Conversely there is a smaller proportion of the work force employed in very large units. For example, only 2 per cent of clothing workers and 5 per cent of textile workers are employed in units of over 1,000 workers, whereas 20 per cent of manufacturing workers are employed in this size of establishment. The true level of employment within small firms in this sector may be even greater than these figures indicate. The main reason is that the clothing sector is often cited as having a large number of small production units operating outside the official economy (i.e. in the 'informal economy').

Table 4.2

Size of establishments in textiles, clothing and footwear
compared with manufacturing

Size band	All Manufacturing	Textiles	Clothing and footwear
	Number of employees		
Total	4,848.9	223.5	310.8
1-9	318.1	9.7	29.9
10-9	252.2	8.1	19.3
20-49	491.1	20.3	40
50-99	505.1	31.3	42.5
100-199	661.7	54.3	64.2
200-499	977.2	64.8	84.8
500-999	652.4	24.8	23.6
1000+	990.0	10.2	6.5
	Number of units in each size band		
Total	154,474	5,310	12,515
1-9	103,588	2,989	8,369
10-19	18,268	585	1,406
20-49	15,902	641	1,321
50-99	7,298	441	619
100-199	4,759	392	462
200-499	3,217	215	294
500-999	963	40	38
1000+	479	7	6

Source: Census of Production. Reproduced in Annual Abstract of
Statistics, 1988, Table 6.13.

A precise calculation of the number of companies or businesses employing this work force within the Leicester City boundary is even more difficult to isolate. The Leicester and District Knitting

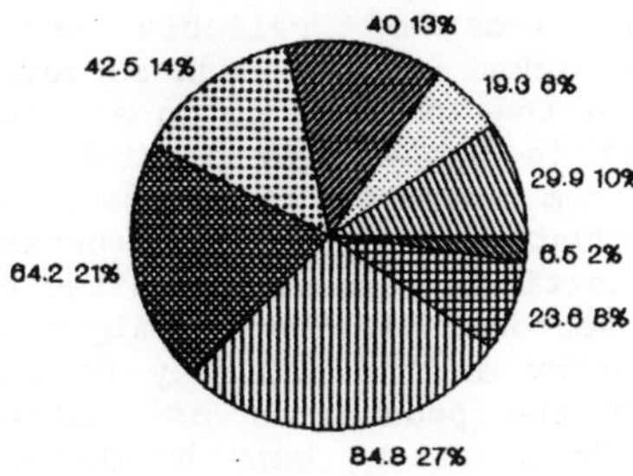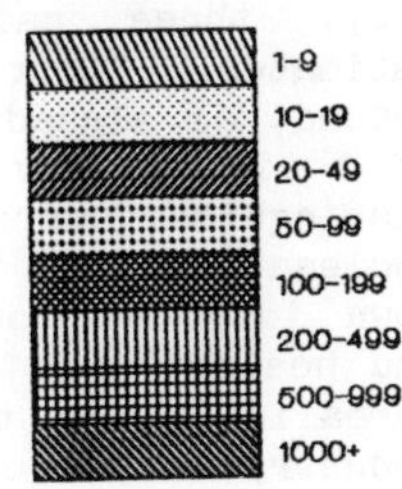

Figure 4.1   Number of employees in each size band – clothing

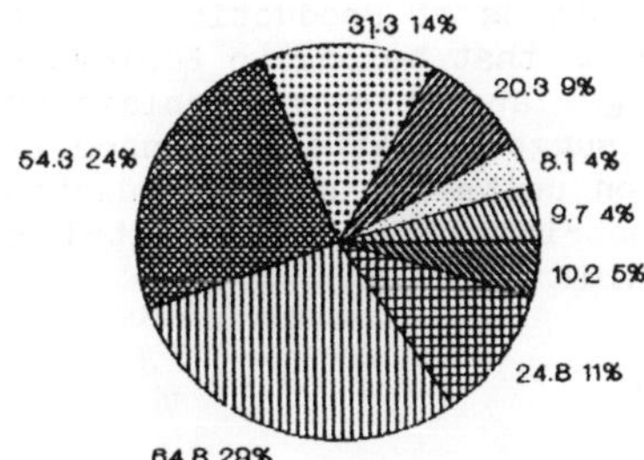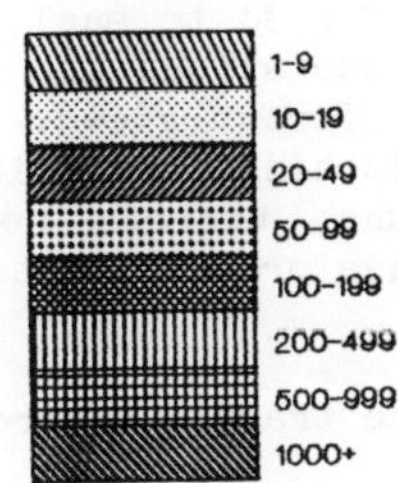

Figure 4.2   Number of employees in each size band – textiles

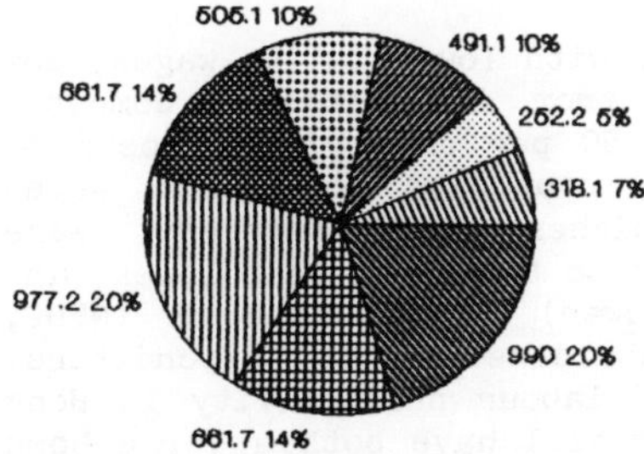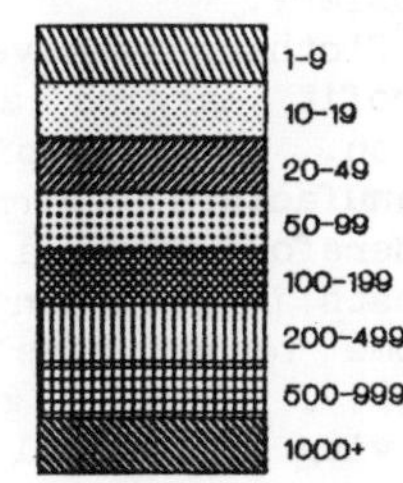

Figure 4.3   Number of employees in each size band – all manufacturing

Industry Association has a membership of 240 knitting companies, but estimates of the number of businesses operating in Leicester could realistically be placed somewhere between 500 and 600 in total (Millington, 1987). Although listings of businesses exist for the city, three main problems arise in terms of reliable total estimates. First, not all companies are picked up by census surveys; second, as already mentioned, companies in the industry go in and out of business very rapidly; and third, it is not always possible to distinguish manufacturing companies from export businesses or wholesale firms. While the national industry is customarily broken down into its four main sectors (i.e. textiles, clothing, knitwear, and hosiery), it is claimed that such a sectoral distinction might be unrealistic for the Leicester-based industry as the majority of the industry in Leicester is involved with the production of knitted garments or fabrics (Parkinson, 1987). In practice, many businesses that claim to work within the 'clothing' sector are in fact mostly concerned with the production of knitted garments (Loman, 1987).

The majority of companies operating in the Leicester City area are of small to medium size, with a work force of anything from two to three employees up to around 200 people. One estimate indicates an average firm size around 70 employees (Boggon, 1987). Our own work and the experience of Leicester City Council suggests a smaller average size of firm in Leicester with a heavy concentration of very small firms (as is also indicated by the Census of Production figures referred to earlier in this section). Firms that belong to Employers' Associations are on average typically larger and this may explain the different views. Leigh and North (1983) suggest that the UK norm for clothing production is for small production units and short production runs, whereas newly industrializing countries have concentrated on mass production techniques in order to minimize costs.

The trading environment of the industry

The textile and clothing industry is now the fourth largest in the UK, employing around 540,000 people and having an annual turnover of £11,700 million in 1985. Only engineering, food, drink and tobacco, and the chemicals industry are larger (National Textile Industry, 1986). The industry is concentrated in the East Midlands, Yorkshire and Lancashire, the South East (primarily London), Northern Ireland and parts of Scotland. The industry is traditionally split into four prime manufacturing sectors: textiles, clothing, knitwear and hosiery.

Clothing is a very competitive industry with low relative wages, low profit margins and continual pressure from international competition. Wages typically contribute about 50 per cent of the costs of manufacture (Birnbaun, 1981). Those countries with low wages therefore have a cost advantage over higher wage countries. Wage costs in Hong Kong in 1981 were estimated to be 50 per cent lower (and some lesser developed countries even lower) than in the UK (NEDC, 1981). Foreign governments have also subsidized capital expenditure, giving additional advantages in terms of labour productivity in Hong Kong, Singapore and Taiwan. India and Brazil have both a large home market and low labour costs which help to make them internationally competitive. They compete in the middle-upper quality product ranges. The UK also faces severe competition from Europe. Italy is said to have a large black economy in clothing which reduces production costs. This, together with their accepted flair for design, makes them formidable competitors. West Germany has high levels of labour productivity and, although wage rates are relatively high, it still manages to compete successfully in a number of areas including quality sports and leisure wear and particularly in menswear.

The textile industry consists of larger enterprises and establishments where long production runs are the norm. This sector is also subject to competition from overseas, particularly the Far East. Newly industrializing countries frequently engage in the production of textiles which can, at least initially, be undertaken with low levels of technology. Government subsidies then allow the industry to introduce new capital and, together with relatively low labour costs, become internationally competitive. The UK industry has been forced to adopt more technologically advanced processes, using more sophisticated machinery and producing materials from man-made fibres. Large parts of the industry have been unable to compete successfully, with the consequent closure of plants and a loss of jobs, output, and share of the world market.

The various subsectors of the textile industry have performed diversely, with woollens and worsteds, hosiery and other knitted goods and textile finishing all recovering to their 1980 levels by 1986 after a sharp drop during 1980/81. Carpets and other textile floor coverings produced 9 per cent above their 1980 level in 1986, while spinning and weaving were some 20 per cent lower (IER, 1987). In textiles the import penetration ratio has increased more rapidly than in clothing and footwear, while the export-sales ratio has only grown slowly, as it has in clothing and footwear (see Table 4.3 and Figure 4.4). The import penetration ratio in textiles increased from 26 per cent to 45 per cent between 1976 and 1986, while the export-sales ratio rose from 25 to 30 per cent. Purchases of plant and machinery for both textile production and for clothing and footwear have increased rapidly over recent years after the setback in 1980.

The Multi-Fibre Agreement has provided some regulations relating to international competition, but import penetration has continued to grow. The import penetration ratio (imports/home demand) grew from 25 to 36 per cent in clothing and footwear between 1976 and 1986 (see Table 4.3 and Figure 4.4). The export-sales ratio grew slightly from 15 per cent to 18 per cent during the same period but the gap between the two widened.

Having failed to maintain its share of consumer spending in the 1970s, the UK textile and clothing industry was operating under the difficult conditions occasioned, on the one hand, by a static home population with a finite spending capacity and, on the other hand, by increasing competition from abroad (Knitting Sector Working Party, 1983). Since that report, however, consumer expenditure on clothing and footwear has grown rapidly (see Table 4.4 and Figure 4.5) but the sector's share of total consumer expenditure has continued to fall. The textile and clothing industry's contribution to the national economy stood at £4,830 million in 1986, a figure which reflects the industry's key role in the nation's economy. With steadily rising export sales reaching £2,870 million in 1985 (£78 million of which was for knitted fabrics and £705 million for knitted clothing), the industry now exports the sixth largest group of UK manufactured products and had a fixed capital expenditure of some £1,300 million in the last few years. The figures for 1986, however, also reveal a widening of the trade deficit to £1,311 million, with imports in textiles increasing at a greater rate than exports (Textile Horizons, December 1986).

Table 4.3
Import penetration ratios and export-sales
ratios in textiles, and clothing and footwear

Year	Import penetration ratio (Import/Home demand)	Export-sales ratio (Exports/Manufacturers sales)
	%	%
Textiles		
1976	26	25
1977	27	27
1978	31	27
1979	33	27
1980	34	30
1981	39	30
1982	39	29
1983	41	28
1984	44	30
1985	44	31
1986	45	30
Clothing and footwear		
1976	25	15
1977	25	19
1978	26	18
1979	29	18
1980	29	19
1981	33	20
1982	33	18
1983	33	18
1984	36	18
1985	35	19
1986	36	18

Source: Department of Trade and Industry, reproduced in Annual
Abstract of Statistics, 1988, Table 12.2.

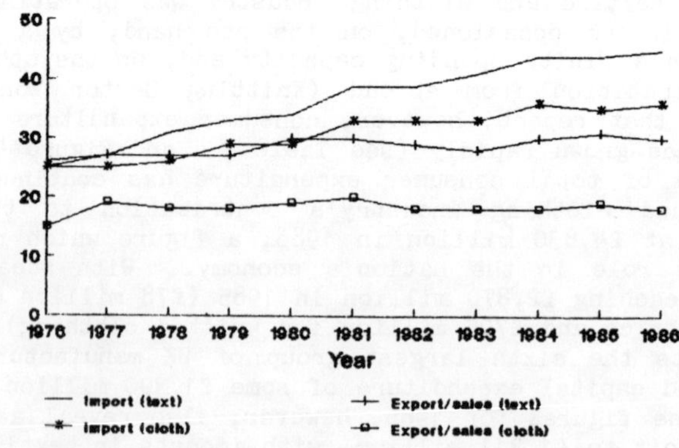

Figure 4.4   Imports and exports in the textile and clothing industry

Table 4.4

Expenditure on clothing and footwear at 1980 prices

Year	Expenditure (£millions)	Share of total consumer spending %
1978	7,596	7.64
1979	8,149	6.91
1980	8,103	5.91
1981	8,105	5.28
1982	8,333	4.95
1983	8,959	4.87
1984	9,497	4.82
1985	10,212	4.75
1986	11,046	4.67
1987	11,857	4.61

Source: Monthly Digest of Statistics, Table 1.5, January 1985,
         June 1986 and March 1988.

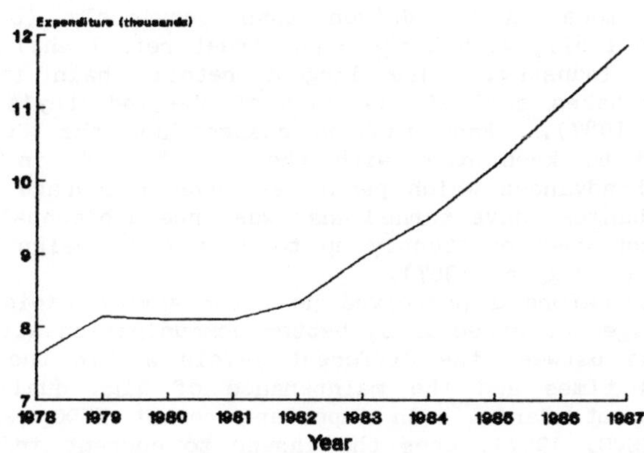

Figure 4.5   Consumer expenditure on clothing and textiles

Textile and clothing figures show a rise in textile imports of 4 per
cent for 1986 with a comparable rise in exports of only 1 per cent.
Similarly, clothing imports rose in the same year by 14 per cent,
while exports of clothing rose by only 4 per cent (British Textile
Confederation, 1987).   A breakdown of clothing imports reveals rises
in sweater, underwear, knitted shirts and hosiery products.   The
steepest rises in sweater imports emanate from the Far East, Portugal
and Greece, with imports from Italy falling.   Total imports of
underwear rose by 11 per cent to 259.8 million garments in 1986, 40
per cent coming from developing countries.   Imports of knitted shirts
increased by 10 per cent, and imports of socks rose by 33 per cent.
Imports of textiles and clothing under the multi-fibre arrangement

31

rose in 1986 by 21 per cent, highlighting the importance of internationally agreed quota arrangements. Potential United States
legislation limiting imports can be expected to harm the UK export
trade further and add to the industry's trade imbalance.

The Textiles Statistics Bureau figures for the years from 1979 to
1985 graphically illustrate the situation facing the UK textile and
clothing industry. While all exports, including fabrics, carpets and
clothing, rose steadily from a 1979 figure of £2,074 million to £3,112
million in 1985, imports over the same period more than doubled from
£2,452 million in 1979 to £5,283 million in 1985. Cheap imports from
the Far East have long been identified as a problem for the UK
industry, but recent figures for trade with the EC support the
industry's concern about the new threat emanating from the poorer
European countries, notably Portugal and Greece. While exports of
textiles and clothing to the EC increased from £903 million in 1979 to
£1,424 million in 1985, imports from the EC rose much faster from
£1,094 million in 1979 to £2,683 million in 1985 (The Textiles
Statistics Bureau, 1986).

For an industry operating in an increasingly international setting
where countries compete for a share of the world's second most
important economic activity, the future health of the UK industry is
seen by some commentators to depend on the retention of internationally agreed import and export quotas coupled with ongoing
investment in high technology, the expansion of exports and the
development of a swift response to home market demands. In a highly
competitive environment, companies in the industry are seen to be
increasingly more market driven than previously (British Textile
Confederation 1987), with large High Street retail chains changing the
face of the industry. The largest retail chain in the UK, for
example, now takes some 20 per cent of the industry's entire output
(Millington, 1987). Fast fashion changes and the whims of younger
buyers eager to keep pace with the newest look in clothing, and
technological advances which permit and even encourage quick response
to fashion changes, have turned what was once a biannual industry into
one which must keep constantly up to date with design initiatives to
achieve orders (Boggon, 1987).

There is therefore a perceived need for a more rapid response with
improved management structures, better communication within individual
companies and between the different levels within the industry, for
shorter lead times and the maintenance of high quality design and
manufacturing standards. An important recent NEDO report, 'Dynamic
Response' (NEDO, 1987), sees the answer to current industry problems
as lying in the speed of response to retailer and market demands,
especially in terms of cutting by half the overall length of the
supply pipeline. However, the point is made that a 'dynamic response'
is all very well but with an apparent drop in home consumption in 1985
to 127.5 million pieces, and a projected static home population
lasting into the next century, there can be no guarantee of an
increase in home consumption. What then is the point of increasing
overall production of a wider range of products at a faster turnaround
rate, unless the consumer can be persuaded to purchase such products?
(Millington, 1987). While there is an obvious requirement for UK
manufacturers to respond adequately to the demands of the retailer-led
home market, the very structure of small companies can prevent such a
response. It is claimed that there is almost no organic gradual
growth from small company size through to large company status (Loman,
1987). Firms will grow from 10 to 20 employees through to 100 or so

over a period of time, mainly serving the needs of the market traders or of small retail outlets. Companies which want to move into the High Street retail-led market have to make a huge jump in size. There would appear to be a barrier at about the 150 employee level, when a company's output will become too big to sustain a market trader level only, but too small to cope with the High Street retailer market level. Only the very specialized company can side-step this problem (Loman 1987). This view of distinct size sectors of the market, coupled with barriers to firm movement between them, is consistent with the theory outlined in section 3 above (see Bosworth, 1987).

Increasing demand for a wider variety of materials and garments has changed the environment for the textile and clothing industry and has brought certain implications to both the retail and manufacturing ends of the trade. New computerized technology has opened up the overall design potential of the industry which in itself has stimulated market demand for new and different products. The greater the variety of outputs being handled by the market the more difficult it becomes for the retailers to plan long term. Retailer uncertainty regarding future requirements makes it difficult for manufacturers to assess their future need for supplies. Shortage of manufacturer supplies leads to inability to meet retailer demands, which results inevitably in loss of manufacturing orders. While the new technology has undoubtedly benefitted those firms able to take advantage of the facilities which it offers, design-initiating manufacturers are increasingly disadvantaged by the ease with which the new technology can copy new designs. Firms investing time and money on 'in house' designers are understandably concerned that such high cost investment should not be poached by other companies who are thus able to avoid a comparable investment (Nathan, 1986).

While it is still an unpalatable fact that companies in the Far East appear able to produce low quality, mass market products far more cheaply than UK manufacturers, there is hope for the home producers insofar as the quality, classical or high fashion products required for the quality stores, provide opportunities for the UK industry. High Street retail buyers find it easier to make contact with home-based manufacturers for supplies of products which need to reflect rapid fashion changes, and so these need to be quickly available. The accent here must be on the maintenance of quality, both in design and garment assembly. The buyers will go abroad for the cheaper, mass produced product. These garments and fabrics are not so prone to fashion change, and the stores can afford to wait a little longer for cheaper supplies to reach them. There are signs that the home market demand for quality products is increasing, and specialist stores catering for up-market clients are proliferating. While our designs are usually believed to be inferior to their European counterparts, better pattern and colour standards and superior make-up are encouraging home market support. Quality design imports from Europe have raised the expectations of the home market, and UK design talent will need to be encouraged and harnessed if the European designer is to be confronted.

Trade barriers erected by some countries still present problems for the exporter and potential import restrictions in the United States may yet become an actuality, but some commentators argue that there are strong grounds for a drive to increase UK exports at the present time. The price gap between the Far East and the UK has been gradually narrowing. The UK industry now has a sounder financial base and is as technologically advanced as many other countries. UK designers are good, though they still lag behind the best in France and Italy, and our technicians are among the best to be found. State aid is available for UK representation at foreign exhibitions and trade fairs in the form of subsidies, help with making contacts, specialist advice on applicable markets and on selling strategies. The question therefore arises as to why exports are not increasing at a greater rate? The answer, for some, lies in the fact that apart from the high quality Scottish trade the UK industry has always been home based and thus foreign buyers are generally unaware of the UK manufacturing and design potential. There is a perceived need for the UK industry to advertize itself more positively and widely than in the past (Millington, 1987). Some of the conclusions of our own research point in this direction as we show in later chapters.

Although much may be done to increase export potential, UK companies, particularly the small ones, appear to face problems when confronting the export market (Boggon, 1987). Financial and time constraints may inhibit the desire to explore export possibilities within the smaller firm and for many UK manufacturers making and maintaining contact with foreign buyers raises difficulties. Advertizing and exhibiting abroad is expensive and the crucial decision as to which export market to choose could present a dilemma. The single greatest problem for the small company, however, is seen to lie in the complexity of form filling and in the rules existing over export tariffs (Boggon, 1987). This situation will change in 1992 when the EC becomes a unified trading area without administrative barriers. This may not be viewed as entirely beneficial to UK firms since it will also open the UK market more fully to European firms. It may also mean that children's clothing will carry VAT and, in the short term at least, this may inhibit sales in the UK.

Textile and clothing output only increased slowly during the 1950s, 1960s and early 1970s. From 1973 to 1981/82 output was on a declining trend with sharp falls in 1974 and 1980. Since 1982 output has increased, though in recent years it has levelled off slightly (see Table 4.5 and Figures 4.6, 4.7 and 4.8). Output increased at an average of 1 per cent per annum between 1980 and 1986, but the IER forecast is for output to remain static up to 1995. Clothing and footwear output increased slowly between 1960 and 1980 and fell sharply between 1979 and 1981. Output has since recovered, though in recent years it has remained static. Output is forecast to grow only slowly up to 1995. Textiles output fell in the 1970s and the subsequent recovery has been smaller than in clothing and footwear. A slower rate of growth is also forecast up to 1990, after which a slight downturn to 1995 is predicted (IER, 1987).

Productivity in clothing and footwear and textiles has increased erratically in recent years. Between 1954 and 1975 the average rate of growth of productivity was 2.5 per cent per annum, although most of the increase came in the early 1970s. The late 1970s saw very little change but, following the 1980/81 recession, the period 1980 to 1986 has been one of high productivity growth averaging 5 per cent per annum. Very similar patterns of productivity increase were apparent in the two subsectors. Productivity growth is predicted to continue at around 3 per cent per annum between 1986 and 1995, with a higher rate of growth in the early part of the forecast period (IER, 1987).

The financial performance of the clothing and hosiery and knitwear sectors of the economy are illustrated in Table 4.6. In 1985/6 profitability in both sectors was healthy, with a 17.5 per cent return

Table 4.5

Indices of textile and clothing industry production

1980 = 100

Year	All textiles	Hosiery and other knitted goods	All clothing
1980	100	100	100
1981	91.8	96.4	92.4
1982	89.5	94.6	93.9
1983	91.6	95.2	96.9
1984	94.1	96.9	103.6
1985	98.4	99.4	110.6
1986	98.9	99.0	112.2
1987	102.9	102.4	110.6

Source: Monthly Digest of Statistics, Table 11.1, June 1986 and
March 1988.

Table 4.6

Financial performance in clothing and hosiery and knitwear

(1985/6)

	Hosiery and Knitwear £000s	Clothing £000s
Sales	2,789,287	2,153,298
Net profit before tax	195,741	128,301
Interest paid	55,207	31,125
Depreciation	63,113	46,092
Capital employed	1,120,884	703,106
Profitability ratios	%	%
Return on capital	17.5	18.2
Return on assets	10.5	10.5
Return on shareholders fund	25.5	23.7
Pre-tax profit margin	7.0	6.0
Gearing ratios		
Borrowing ratios	49.9	52.8
Equity gearing	0.4	0.4
Income gearing	22.0	19.5

Source: Industrial Performance Analysis 1987/8, ICC.

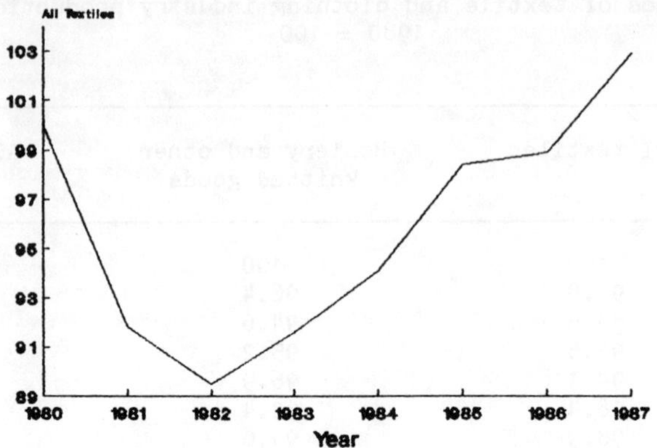

Figure 4.6   Index of textile production

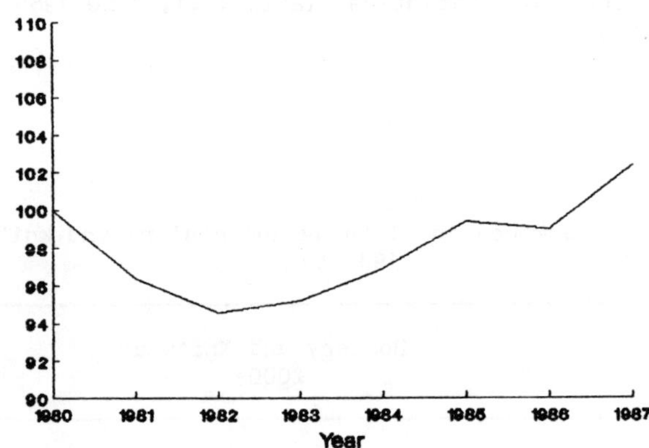

Figure 4.7   Index of hosiery production

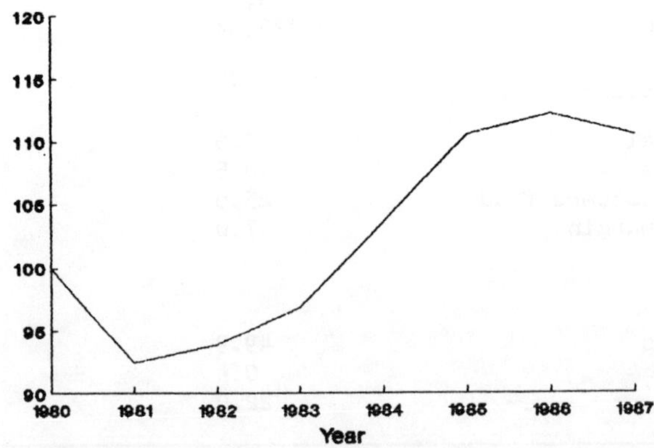

Figure 4.8   Index of clothing production

36

on capital in hosiery and knitwear and 18.2 per cent return on clothing. Table 4.7 and Figures 4.9 and 4.10 show producer prices for both textiles, and clothing and footwear in recent years. In textiles, output prices are around 48.5 per cent higher now (early 1988) than they were in 1980, while input prices for materials and fuel are 46.8 per cent higher. In the early 1980s, however, input prices were rising substantially faster than output prices causing a squeeze on profits, which has been marginally reversed since 1985. In clothing and footwear, however, input prices of materials and fuels are some 49 per cent higher than in 1980, while output prices are only 40 per cent higher, squeezing profits in this sector. In clothing and footwear the gap between input price inflation and output price inflation in the early 1980s widened only gradually. In the years from 1984 the gap has remained, though not widened, so that manufacturers have been able to cover rising input prices by passing them on to consumers but they have been unable to recoup earlier losses.

Table 4.7
Index of producer prices for inputs and outputs
for textiles, and clothing and footwear
1980=100

Year	Price Index Materials and Fuel Purchased	Price Index Output (Home Sales)
Textiles		
1980	100.0	100.0
1981	105.8	104.5
1982	114.8	110.4
1983	123.7	116.4
1984	136.6	123.1
1985	139.8	129.6
1986	133.8	135.7
1987	142.1	142.1
1988	146.8	148.5
Clothing and footwear		
1980	100.0	100.0
1981	102.4	103.8
1092	109.9	107.9
1983	117.4	112.5
1984	129.2	118.2
1985	136.9	125.4
1986	139.1	130.8
1987	145.2	135.3
1988	149.1	139.9

Source: Monthly Digest of Statistics, Table 18.6, March 1988.

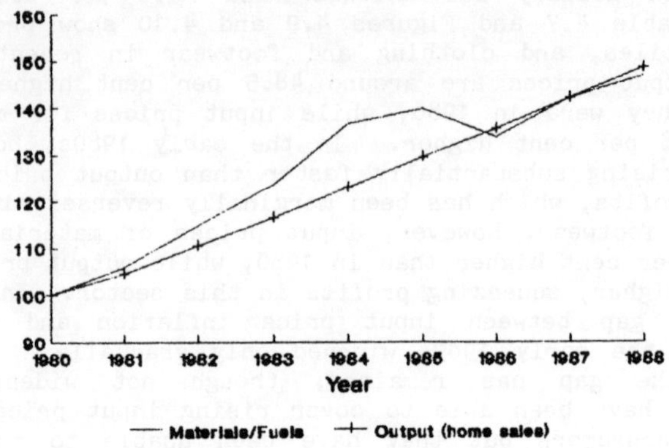

Figure 4.9   Index of producer prices, textiles

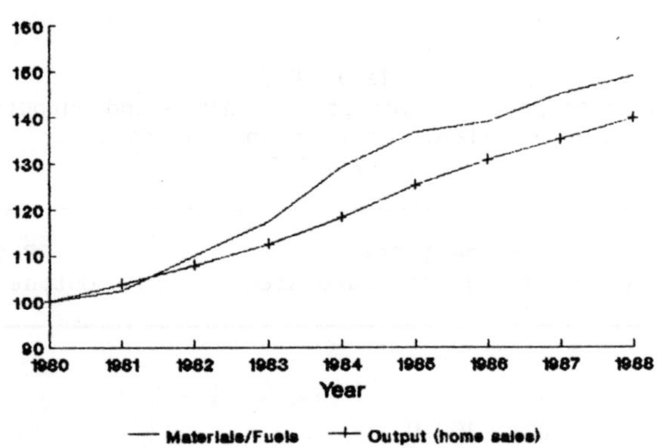

Figure 4.10   Index of producer prices, footwear and clothing

The labour force of the industry

The very nature of the textile and clothing industry in the UK makes it virtually impossible to arrive at finite employment numbers either at the national or regional levels. Disproportionately high liquidation rates, estimated at some 85 per cent of all UK voluntary liquidations and 27 per cent of all compulsory ones, reveal a situation where numerous, often small, businesses move repeatedly into and out of operation with an accompanying shifting of labour into and out of work (Totterdill and Pearce, 1986). While there are some very large manufacturing companies in the industry, the majority of businesses operating in the textile and clothing trades are small or very small, are generally family owned and staffed, and often utilize the services of outworkers who may or may not be registered as employees. Ethnic owned businesses have sometimes been found to be particularly difficult to reach as they may often prefer to work within their cultural boundaries which impinge little on the established white sector of the industry (Boggon, 1987).

While estimates vary widely, industry employment figures exist which identify the textile and clothing industry as being a prime employer of labour within the UK. Estimates from the Department of Employment for June 1987 reveal a total of 432,100 people employed in textile and clothing production (222,400 in textiles and 209,700 in clothing) in Great Britain, and this total accounts for over 6 per cent of employment in all production industries. Comparative 1987 employment figures for other major employing industries show the chemical industry employing 336,000; the motor vehicles industry employing some 240,000; the coal mining industry 154,000; and the aerospace industry with 158,000 in employment (<u>Employment Gazette</u>, November 1987).

The industry has only a small self employed sector. In textiles only 2.5 per cent and in clothing and footwear only 5 per cent of the total work force were self employed in 1986. About 9 per cent of the work force worked part time. Employment in textiles is shared almost equally between males and females. Employment in clothing and footwear is dominated by females who occupy 70 per cent of the jobs. There are marked differences between the structure of male and female employment within both sectors (see Table 4.8 and Figures 4.11 and 4.12). Taking the group as a whole, 23 per cent of male workers were in managerial or professional and related occupations in 1986 compared with only 8.5 per cent of female workers; 72 per cent of females and 52.5 per cent of males were operatives and labourers; 18.5 per cent of males and only 4.5 per cent of females occupied craft and skilled occupations; finally 6 per cent of males and 15 per cent of females were in clerical and secretarial or sales and personal service occupations (IER, 1987).

Table 4.8
Occupational structure in textiles and clothing

Occupation	Males and Females
	%
Operatives and labourers	65
Craft and skilled	10
Sales and personal services	4
Clerical and secretarial	8
Professional and related	5
Managers	8

Occupation	Males	Females
	%	%
Operatives and labourers	52.5	72.0
Craft and skilled	18.5	4.5
Sales, personal services, clerical and secretarial	6.0	15.0
Managers and professional and related	23.0	8.5

Source: IER Review of the Economy and Employment, 1987.

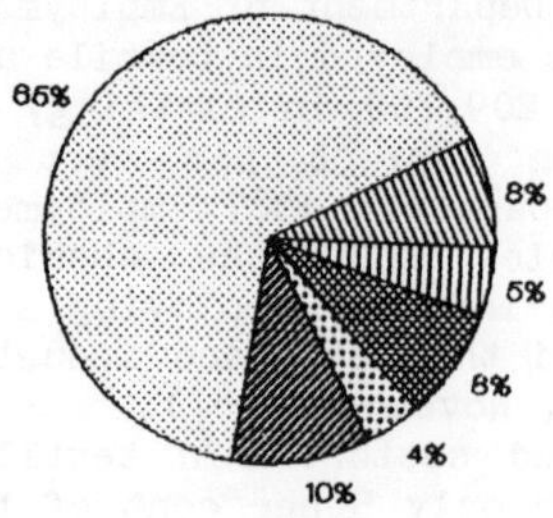
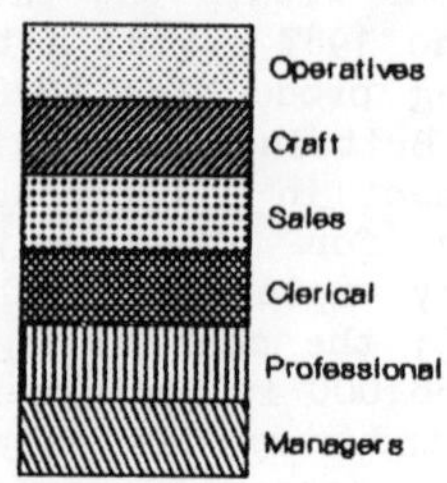

Figure 4.11   Occupational structure in textiles and clothing

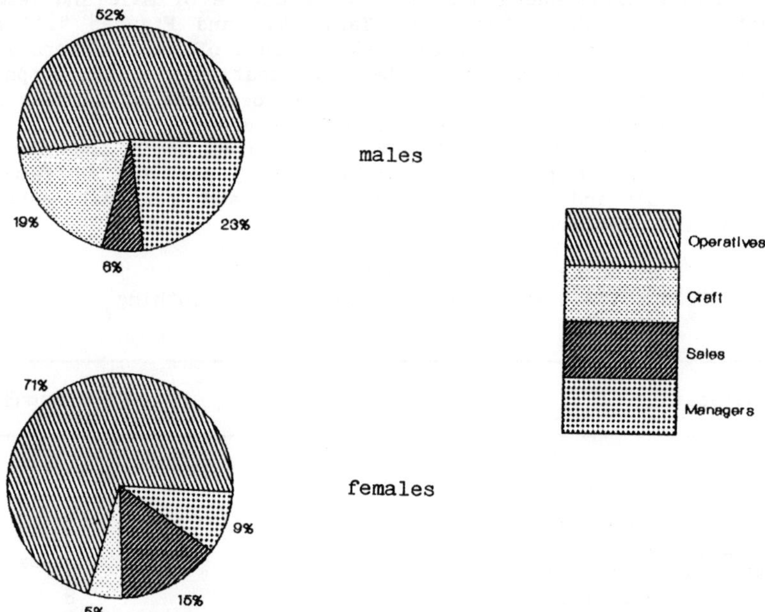

Figure 4.12   Occupational structure in textiles and clothing,
                males and females

Industry employment estimates for the Leicester City textile and clothing trades varies between 17,000 and 25,000 persons, the majority of whom are women. The National Union of Hosiery and Knitwear Workers claims some 8,500 members and estimates that as many people again are non-unionized (Loman, 1987).   The Leicester and District Knitting Industry Association estimates that some 25,000 people are working within the industry in the Leicester City area (Boggon, 1987).   The 1981 Census identifies between 25,000 and 26,000 textile and clothing workers, with approximately 16,000 employed in the knitwear sector.

Estimates of the gender and ethnic characteristics of the Leicester textile and clothing industry labour force were given by the National Union of Hosiery and Knitwear Workers, the vast majority of whom are engaged in the knitwear trade (Loman, 1987). These figures show that between 70 and 75 per cent are female of either white or Asian ethnic orientation. (This percentage rate equates with national percentages). While there is no ethnic percentage breakdown for the Leicester industry, it is estimated that Asian women appear to be proportionately represented within the general work force, while Asian men are not. This situation, it is suggested, is rooted in an historical ethos based on gender segregation of work, a tradition which persists today. Women are traditionally employed in the pressing, cutting, making up and packaging stages of the manufacturing process. Asian women are well represented in these aspects of the trade. There appears to have been few barriers to their absorption into the UK industry at these work related levels. The gender work orientation for men in the UK textile and clothing industry has been somewhat different. Men have historically been employed as knitters, mechanics, dyers and finishers. These skills have often been handed down from father to son within the white labour force, who have operated an informal 'closed shop' policy in order to protect their position within the industry. Asian men may, therefore, have found it more difficult to break into the UK industry than have their female counterparts. Asian men continue to be disproportionately under-represented within the Leicester-based textile and clothing industry in general (Loman, 1987).

While the textile and clothing trade remains the fourth largest industry in the UK at the present time, employment figures for recent years show a marked shrinkage in terms of people employed. In 1971, slightly over one million people were employed throughout Great Britain in the industry. This figure had shrunk to 520,000 by 1988 (see Table 4.9 and Figure 4.13). All sectors of the industry have experienced job losses. It is anticipated that total employment will reduce further in the future. In its latest review the Institute for Employment Research projects that a further 50,000 male jobs and 60,000 female jobs will disappear between 1986 and 1995. For males there are no expanding areas. Female jobs are predicted to increase for certain occupations, mainly among professional and related occupations. The main area of job losses for both males and females will be in the operatives and labourers category, where 33,000 jobs are predicted to disappear for males and 47,000 for females. Operatives and labourers currently account for 65 per cent of the work force in the industry, 10 per cent are craft and skilled workers, 8 per cent are managers, 8 per cent clerical and secretarial, 5 per cent professional and related and 4 per cent sales and personal services (IER, 1987).

One reason for the predicted further job losses may well be the increasing use of computerized technology which is now undertaking the work that was previously exclusively manual (Harrison, 1987). Although investment in up-to-date technology is seen as being of paramount importance to the future health of the industry, for some a continuing investment in people remains essential with future success dependent on the maintenance of a well trained and committed work force from shop floor to top management (Knitting Sector Working Party, 1983). The textile and clothing industry is generally regarded as having good labour relations, with unions which understand the problems facing the industry. However, these same unions are now expressing worries about the working conditions of their members. They identify a growing erosion of the piecework payment structure, with companies increasingly preferring to pay by time. It has been argued that this trend will lead to declining pay levels in what is already perceived to be a generally low paid industry (Loman, 1987).

Table 4.9
Employees in employment in Great Britain in
textiles, leather, footwear and clothing
(divisions of SIC 1980 43-45)

Year	Employees (000s)
1971	1,010
1972	986
1973	758
1974	769
1975	731
1976	720
1977	719
1978	712
1979	713
1980	705
1981	664
1982	638
1983	599
1984	582
1985	573
1986	552
1987	531
1988	520

Sources: Employment Gazette Historical Supplement No. 1,
            February 1987, Volume 95 No. 2,
            Employment Gazette, May 1988.

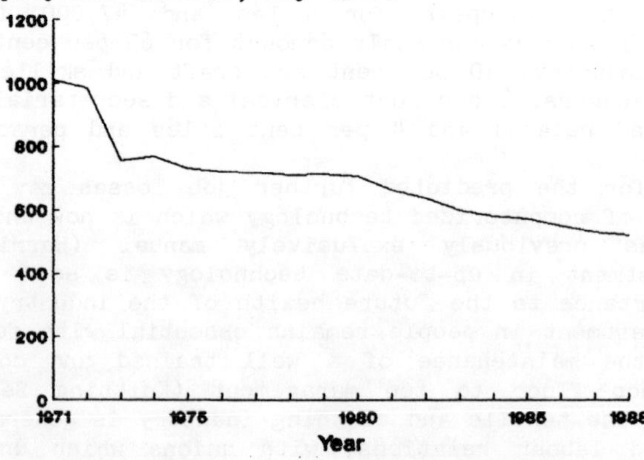

Figure 4.13   Employees in employment (divisions 43-45)

The pay of workers in the industry relative to other workers in manufacturing and in the economy as a whole is shown in Table 4.10 from the latest New Earnings Survey. These figures indicate that male manual workers in the textile industry receive average gross weekly earnings 11 per cent lower than all male manual workers in all industries and services and some 15 per cent lower than those in manufacturing. Male manual workers in clothing receive 12 per cent less than in all industries and services and some 17 per cent below manufacturing wages. Female manual workers in these industries also receive lower average wages; in textiles these are 11 per cent lower than in all industries and services, although only 4 per cent lower than those for females in manufacturing; in clothing, average weekly earnings for females are 15 per cent lower than in all industries and services and 9 per cent lower than in manufacturing industries alone.

Table 4.10
Earnings in textile and clothing industries compared
with manufacturing and all industries and services
(excluding those whose pay was affected by absence, hourly rate
excludes payment for overtime)

	Average gross weekly earnings	Average hourly	Total	Normal basic	Overtime
Full-time manual males on adult rates					
All industries and services	185.5	404.3	44.6	39.1	5.5
All manufacturing	195.9	423.8	44.7	39.0	5.7
Textiles	165.6	353.3	45.6	39.3	6.3
of which: Hosiery and other knitted goods	166	381.6	43.2	39.3	3.9
Clothing, footwear and leather	162.7	386.7	41.9	39.8	2.1
of which: Clothing, hats and gloves	136.2	323.3	41.6	39.1	2.6
Full-time manual females on adult rates					
All industries and services	119.7	292.4	40.2	38.5	1.8
All manufacturing	110.9	282.2	39.0	37.5	1.4
Textiles	106.7	265.9	39.8	38.5	1.3
of which: Hosiery and other knitted goods	104.9	266.6	39.2	38.4	0.8
Clothing, footwear and leather	101.3	256.5	39.2	38.4	0.8
of which: Clothing, hats and gloves	99.2	253.3	39.0	38.2	0.8

Source: New Earnings Survey, 1987.

The increasing adoption of expensive high technology within the industry is causing management to call for the labour force to work what might be described as 'unsocial hours'. Shiftworking is increasing and some of the labour force are already working Saturday shifts. Unions believe they will have to concede to Sunday working in the future. Companies which invest in costly computerized machinery often want a twenty-four hour day, seven-day week operational capacity (Loman, 1987). Such developments may be viewed as bringing declining working conditions for the labour force operating in the industry. However, in principle, new technology can also bring benefits to firms and workers. Routine work can often be undertaken by machines to release workers for more interesting tasks. Firms which invest in new technology to improve productivity and reduce unit costs become more competitive and are more likely to gain orders. Additionally those firms may also achieve increases in profits which are frequently passed on to some extent to the work force in any number of ways, from increases in rates of pay or bonuses to improvements in the working environment.

Skill shortages are a major problem perceived by management, unions and employers' associations alike. While it is felt to be true that new computerized equipment is gradually de-skilling the work force, nevertheless, skilled people are still required for the industry (Teague, 1987). Within the textile and clothing industry shortages are caused partly by lack of investment in training and partly by the recent recession. When the recession hit the industry in the early 1980s, skilled people left to go into other types of work (mainly shop work) and many older workers retired. The pool of skilled labour gradually drained away. At the same time, the new Industrial Training Act allowed employers to opt out of their compulsory training require-ments. Employers chose to set up their own voluntary training body rather than support Industry Training Board programmes. The new training body employs only two people and the training of skills therefore has greatly diminished. Large companies have been forced to maintain their own expensive training schools and feel that they are subsidizing the local industry by supplying skilled labour (Loman, 1987). The YTS scheme has reversed the non-training trend somewhat, as this scheme carries a compulsory training component. The problems of finding skilled workers are also associated partly with the bad image of the industry which deters young, talented people from entering, and partly with out-dated management methods which need to be upgraded (Boggon, 1987). Leicester City Council is working with the Clothing Industry Training Board (CAPITB) on several training initiatives. The Leicester Knitwear Training Group (LKTG) is also active in training and managed by the Leicester knitting industry through LADKIA.

The continuing shortage of skilled labour within the industry poses what has been described as one of the most serious problems for its future. To reverse this trend, management will need to appreciate the need for the training of a versatile and flexible labour force which can adapt to new technological developments and market demands. Increased support from the industry for sandwich design courses might tempt new, young talent into the textile and clothing trades, thus strengthening the existing pool of quality designers with which to confront home market demands and export potential. There would appear to be a need for effective educational representation on local industrial training committees (Knitting Sector Working Party, 1983). There is evidence that the subject of textiles is still taught in schools as an 'art and craft' subject and not as a technological one. Often it stands as an option for the less academically 'bright' student. To change this situation, textiles could be taught in its own right in schools, and as a technology could be an integral part of a science and technology syllabus. Courses in textile technology should be available for teachers and new exams should be offered to students, especially in textile design for the 'high flyers' (Millington, 1987).

Industry investment in labour force training currently stands at 2.3 per cent of annual sales turnover but much of this investment is being negated by a high level of staff turnover, particularly in respect of new, young recruits to the industry. If advances in multi-skilling, in handling people, and in liaison with unions are to be achieved, then professional training programmes need to be established which deal with all aspects currently confronting the textile and clothing industry (Cole, 1987). While effective training may be said to begin with efficient recruitment and selection, the efficacy of any structured training for the industry which may be implemented may well rest on the industry's ability to formulate long term career prospects for its work force (Cole, 1987, Loman, 1987). At present, a typical female machinist enters the industry in her early twenties and retires from the industry maybe forty years later - as a machinist. A more open career structure which encourages talent and ambition and offers promotion to management functions promises better for the future.

Any training improvements which may be achieved will, nevertheless, be dependent on management awareness and capability to train effectively (Apparel International, 1987) but, it is claimed, managements do not know how to put together effective training programmes or how to monitor progress, results and costs (Cole, 1987). It is difficult for a sector which has developed as a low skill, low pay industry to attract recruits who are looking for career prospects. It will need at least to negate its former image and probably make real changes before this will happen. The 'hit and miss' situation, as it relates to the training of the work force at both the shop floor and management levels, urgently needs to be addressed if current skill shortages and recruitment difficulties are to be reversed (Parkinson, 1987).

Image of the industry

The UK textile and clothing industry has an image which is scarcely conducive to the recruitment of labour or likely to result in an upgrading within the common consciousness of the nation. Currently its image is one of long hours, poor pay, and bad working conditions. Nor has this image been unfairly accredited in the past. Sweatshop conditions have long been identified as one of the demerits of the industry, and poor working conditions, particularly in respect of health and safety, continue in many quarters of the industry today (Totterdill, and Pearce, 1986, Loman, 1987). Sick pay is practically non-existent in the industry and holiday pay is little better for many workers. Earnings, while good when work is available, is erratic. Company orders cannot be guaranteed and therefore neither can security of employment and income for the labour force (Loman, 1987).

While the advent of the new technology, the improvement in quality of home produced garments, and the rising fashion consciousness particularly among the young, has gone some way towards a revamping of this image, much has still to be done if talented new people are to be attracted into the industry (Boggon, 1987). More harmful, perhaps, than the actuality of poor working conditions or erratic earnings, is the prevailing perception of the home industry as having had its day, and being now in irreversible decline (Loman, 1987).

As already discussed, the introduction and restructuring of career ladders in all parts of the industry might help to stimulate new interest, but there are clearly problems in a structure which often rewards promotion with a reduction of salary because of a loss of piecework rates of pay (Loman, 1987). Investment in new, high technology machinery may instil more confidence and act as a lure, but this same machinery is often perceived as de-skilling the current work force and so may not help in attracting quality labour into the industry. When new technology is introduced, management will frequently be faced with a choice of two routes. One involves the polarizing of skills; de-skilling the majority of jobs while re-skilling certain key jobs which become more technical. The alternative is to enhance skills more widely. This latter option often involves multi-skilling or flexi-skilling of a larger proportion of the work force. Many more workers gain some knowledge of different tasks associated with the new machinery. This usually involves workers having more responsibility for making small changes to the machinery and more control over their output in terms of both quality and quantity. This type of option may not be appropriate in an industry where workers have previously undertaken only very repetitive work operations and may not therefore react favourably to or be trusted with this additional responsibility. There is no doubt however that multi-skilling can enhance the quality of the job, though it can also make additional demands which individuals may be unwilling to bear without sufficient financial rewards.

5 Survey of the Leicester textile and clothing industry

The survey firms

After visiting five firms identified for the purpose of a pilot survey in April and May 1987, a small number of amendments to the questionnaire were made prior to the full survey. On the 16th and 17th June, a total of 350 textiles and clothing trade firms in the Leicester City area were circulated with the research project survey questionnaire. The names and addresses of these firms were drawn from the Leicestershire County Council data bank of manufacturers in the Leicestershire county area. The initial response was good but not overwhelming and the questionnaire was followed up with telephone calls, those firms selected to be followed up being chosen randomly. This process proved extremely time consuming. Few firms were reluctant to complete the questionnaire but many questionnaires did not materialize when they had been promised. Eighty-six completed questionnaires have been used to compile the results which form the body of this report. These were completed by senior personnel within the firms; 29 by a director or chairperson, 26 by the managing director or chief executive, 13 by the proprietor or partner, 5 by the company secretary, and 5 by a sales or finance director. The others were completed by an administrator, a general manager, a personnel officer, a personal assistant to the chairperson, and a company secretary/financial director.

Methodology

The information currently available from survey respondents has been subjected to computer analysis using SPSSx. Results reported here have been obtained using frequency counts, percentage responses, cross tabulations and, in the case of the introduction of new technological systems, some regression analysis.

Survey findings

Characteristics of the survey firms

<u>Age of Firm</u> Respondent information reveals a concentration of new firms in the textile and clothing sector in the Leicester city area during the past 18 years. Of those who gave their precise date of establishment, 47 firms have become established since 1969; 26 firms were established during the 1970s and 21 were established during the 1980s. Of the remaining firms, 31 were established prior to 1969; 6 are very old established having started life pre-1900, and 25 firms were formed between 1900 and 1969. The remaining 8 firms are more than 5 years old but have not given their precise date of formation.

Table 5.1
Age of firm

Period of firm formation	No. of Firms
Pre-1900	6 (7%)
1900-1969	25 (29%)
1970s	26 (30%)
1980s	21 (24%)
More than 5 years old (precise date of formation not known)	8 (9%)

<u>Legal status of firm</u> A very high proportion of respondent firms, 72 per cent (62), have adopted the legal status of the private limited company; 20 per cent (17) are partnerships; 7 per cent (6) are sole proprietorships. (One firm has not responded to this question).

<u>Type of firm</u> The large majority of respondent firms, 86 per cent (74), are single, independent manufacturing units; 9 per cent (8) of firms are part of a larger group of companies. One firm is a part of a smaller group of companies, while another is a division of a private limited company and yet another is a member of a group of independent manufacturing units. (One firm did not respond to this question).

<u>Owner-managed firms</u> Again, a very high proportion of respondent firms, 84 per cent (72), are managed by the owner; 14 per cent (12) are not. (Two firms did not complete this question).

<u>Ethnic origin of owner</u> According to respondent information, 60 per cent (52) of the firms' owners are from a white ethnic origin, and 31 per cent (27) of owners are of Asian ethnic origin. Ten respondents declined to answer this question using a variety of methods to avoid giving this information. The survey has not revealed the existence of any Afro-Caribbean owners operating in the Leicester City area.

Patterns of employment

<u>Size of firm</u> While it seems difficult to achieve general agreement as to what constitutes a small firm, respondent information reveals the

48

existence of a large proportion of small, or relatively small, firms
operating in the Leicester City area. The survey revealed that 83 per
cent (71) of the firms employ fewer than 100 people; 64 per cent (55)
of all firms employ fewer than 50 people; and 36 per cent (31) employ
fewer than 20 people. Of the remaining 17 per cent, 9 employ between
101 and 200 people, 3 between 201 and 500 and 2 firms employ over 500.
(One firm did not reply to the question).

Table 5.2
Size of firm

No. of Employees	No. of firms (%)
Fewer than 20	31 (36%)
Between 20 and 50	24 (28%)
Between 50 and 100	16 (19%)
Between 100 and 200	9 (10%)
Between 200 and 500	3 (3%)
Over 500	2 (2%)
Not known	1 (1%)

<u>Gender of employees</u>   The majority of firms use a preponderance of
female labour: 77 per cent employ a majority, and in 42 per cent of
firms more than three-quarters of workers are women.

<u>Part time employment</u>   Only 24 per cent of firms employ no part time
workers but the majority employ relatively small proportions: 60 per
cent employ 15 per cent or fewer part time workers and only 9.2 per
cent employ over 50 per cent part time workers.

<u>Employment by race</u>   Relatively small numbers of Afro-Caribbeans are
employed.   Only 2 firms employed over 10 per cent Afro-Caribbean; 60
per cent employed none. Without exception, Asian owned firms employ a
majority of Asian workers and white owned firms employ a majority of
white workers: i.e. 27 firms employ a majority of Asian workers; 53
employ a majority of white workers.   Only 17 firms employ all white
workers and 6 employ all Asian. (Six firms did not respond to the
question.)

<u>Use of outworkers</u>   Respondent information suggests that the use of
outworkers continues to feature prominently as an employment option to
firms operating in the Leicester area: 63 per cent (54) of firms
employ outworkers - at least from time to time.   Relatively small
numbers of outworkers are usually employed by firms: 43 per cent (37)
employ 5 or fewer, but 2 firms employ more than 100 outworkers.
Outworkers do not usually represent a majority of the work force, only
1 firm is in this position.   The majority employ fewer than 25 per
cent of their work force as outworkers.

<u>Use of freelance specialists</u>   A little under one-half of respondent
firms employ freelance specialists for various aspects of their design
and manufacturing processes: 42 per cent (36) of firms employ

49

freelance workers, if only occasionally. The greatest need for freelance specialty would appear to be in the field of design, with 19 of
these 36 firms employing freelance designers of some kind. The survey
responses showed that 10 firms employ freelance mechanics and 7 employ
freelance management consultants, mostly for their marketing and
finance skills. Pattern preparation and cutting, fabric graphing,
computer programming and accounts were also mentioned. Firms usually
use only one or two freelance workers, and no firms use more than
five.

Raising finance

<u>Sources of finance</u>   Respondent information reveals that banking
institutions are a major source of finance for the firms: 84 per cent
(72) of the firms state that they use banks as a source of borrowing;
36 per cent (31) utilize retained profits to raise finance; 23 per
cent (20) use family loans as a means of financing the business; 23
per cent (20) utilize share capital; and 17 per cent (15) of the firms
use other sources. These other sources include leasing, the group or
parent company, finance companies, hire purchase, the Small Firms'
Guaranteed Loan Scheme and raising capital against the owner's home.

Table 5.3
Sources of finance

Source of Finance	No. of firms
Banks	72 (84%)
Retained profits	31 (36%)
Family loans	20 (23%)
Share capital	20 (23%)
Other sources	15 (17%)

<u>Problems associated with borrowing requirements</u>   Only 15 firms
report any difficulty in raising capital. Of those who do report
experiencing difficulties, 10 are Asian owned firms. Smaller firms
have more difficulty than large firms. Of the 15 firms who have had
some difficulty: 6 employ fewer than 20; 6 between 20 and 50; 2
between 50 and 100 and 1 between 100 and 200. The most common
explanations of the difficulties that they encountered were associated
with bank caution at long term investment in what was considered to be
a high risk industry, or problems connected with establishing security
against bank loans or overdrafts. One respondent suggested that
certain advertizing campaigns might be said to run counter to the
reality: 'The Action Bank - no action!', 'The Listening Bank won't
listen.'   The difficulties encountered by all firms were directly
associated with banking institution policy and practice.

<u>Turnover of firms</u>   Over half of the respondent firms, 51 per cent
(44), report an increase in their total volume of trade during the
period 1986 to 1987; 41 per cent (35) of firms report that their
volume of trade has remained stable during this same period; only 7
firms report a decrease in their volume of trade. For the current

financial year, 1987/88, 59 firms anticipated an increase in their total volume of trade, while 26 did not. (Nine did not reply to the question).

Products and production methods

<u>Categorization of firms</u> The survey can be said to have drawn a fairly even spread of responses across the 4 main sections of the textiles and clothing sector of industry, 34 firms categorizing themselves as being knitwear firms; 25 clothing firms; 19 hosiery firms; and 15 textile firms. Some firms have placed themselves in more than one category, and 9 have categorized themselves in other ways: trimmers, dyers and finishers, ladies underwear, design and technical services, commission knitters and importers.

<u>Methods of production</u> Respondent information reveals a high proportion of firms who cut and sew fabric into finished goods using either bought-in fabric or fabric produced by the firm: 36 per cent (31) of firms buy in fabric to cut and sew, while 44 per cent (38) produce their own fabric which they also cut and sew. The responses indicated that 17 per cent (15) produce their own knitted fabric to sell on, and another 15 per cent (13) of firms produce hosiery; 15 per cent (13) of firms produce body blanks; 4 firms produce fully fashioned garments, and 2 firms produce part-finished garments. Another 7 firms mentioned other activities: raw yarn to finish, dyeing and finishing, cutting, making and trimming and making only, importing and commission knitting. Firms are often involved in more than one activity.

Table 5.4
Methods of production

Activity	No. of Firms
Bought fabric: cut and sew	31 (36%)
Own fabric: cut and sew	38 (44%)
Own knitted fabric to sell on	15 (17%)
Hosiery	13 (15%)
Body Blanks	13 (15%)
Fully fashioned garments	4 (5%)
Part finished garments	2 (2%)
Other (raw yarn to finish, finishing, cutting, making and trimming, making only, importing, and commission knitting)	7 (8%)

<u>Main product lines</u> Respondent information so far reveals a roughly equal spread throughout the main product lines of the four sections of the industry. Equally represented are knitted outerwear for men, women and children and other outerwear, again for all sections of the market, including sweatshirts, trousers, blouses, jeans and jackets. Underwear, lingerie and nightwear products for all three sections of the market are well represented, as is hosiery - this being defined as

footwear as distinct from underwear - as well as leisure and
sportswear garments. Fabric manufacture, tapes, bindings and trims
are all represented among the responses.

<u>Subcontracting of work</u> About half of the respondent firms, 51 per
cent (44), indicate that they subcontract work out to other firms.

Markets

<u>Outlets used by firms</u> Many of the firms sell their products to more
than one type of outlet. A large proportion, 71 per cent (61), of
firms use wholesalers; 45 per cent (39) sell to large chain stores; 40
per cent (34) sell to single retail outfitters; and 40 per cent (34)
of firms sell to mail order companies. Only 7 per cent (6) of firms
sell through factory shops compared with 21 per cent (18) which sell
through a variety of other outlets including the Health Service, small
chain stores, merchants, exporters, a marketing consortium, brand
names, other producers who then 'sell on', and direct to the general
public.

Table 5.5
Outlets used by firms

Type of Outlet	No. of Firms
Wholesalers	61 (71%)
Large chain stores	39 (45%)
Single retail outfits	34 (40%)
Mail order	34 (40%)
Factory shops	6 (7%)
Other outlets	18 (21%)

<u>Location of outlets</u> The overwhelming majority of respondent firms, 92
per cent (79), sell at least some of their products to outlets located
in parts of Britain other than the East Midlands. Most firms sell to
a variety of locational outlets, but 62 per cent (53) of the firms
sell at least some of their products locally. About 45 per cent (39)
of firms export at least some of their products, and 43 per cent (37)
sell to outlets located in the East Midlands area.

<u>Export destinations</u> The 45 per cent (39) of firms who trade abroad do
so to a variety of different countries, a substantial number of them
trading with more than one country. Europe is the most popular
destination with 85 per cent (33) of exporters sending goods there
compared with 23 per cent (9) of firms exporting to North America, and
33 per cent (13) that export to the rest of the world.

Sources of yarn and fabric

A high proportion of the firms obtain their yarn or fabric from local
suppliers: 69 per cent (59) buy British-made yarn or fabric from local
suppliers, while 64 per cent (55) buy foreign-made yarn or fabric from
local suppliers; 40 per cent (34) of firms buy British-made yarn or

fabric from non-local suppliers, while 29 per cent (25) buy foreign-
made yarn or fabric from non-local suppliers; 33 per cent (28) of
firms buy yarn or fabric from foreign suppliers.

Most firms use both British and foreign yarn and/or fabric. Only 12
firms use exclusively British yarn/fabric, while 13 use exclusively
foreign-made yarn/fabric. Similarly 69 firms use at least some
British yarn/fabric and 70 firms use at least some foreign yarn/
fabric.

Table 5.6
Sources of yarn and fabric

Origin of Yarn or Fabric	No. of firms
British-made yarn/fabric only	12
Foreign-made yarn/fabric only	13
Both	57
Locality of suppliers	
British yarn/fabric from local suppliers	59
British yarn/fabric from non-local suppliers	34
Foreign yarn/fabric from local suppliers	55
Foreign yarn/fabric from non-local suppliers	25
Foreign yarn/fabric from foreign suppliers	28

New technologies in the Leicester textiles and clothing industry

Competitiveness and efficiency

Respondent information suggests a gradual scaling down of self
assessment on competitiveness and efficiency compared to the proximity
of other companies in the field. For example, while 85 per cent (73)
of firms believe that they are as competitive as other locally-based
firms, slightly fewer firms, 69 per cent (59), believe they are as
competitive as firms based in other parts of Britain. To complete
this scaling down of self assessment, only 31 per cent (27) of firms
believe they are as competitive as foreign firms.

A similar pattern is revealed when respondents self assess their
technical efficiency in relation to competitors: 74 per cent (64)
believe they are as efficient as locally-based competitors; 62 per
cent (53) believe they are as efficient as firms in other parts of
Britain; but only 41 per cent (35) believe they are as efficient as
foreign firms.

This pattern may be attributable, at least in part, to an increasing
lack of participant knowledge of competitors based outside the local
area and abroad, as evidenced by the relatively high proportion of
firms who did not respond or replied 'don't know' to these two sets of
questions. For example, while 15 per cent of firms failed to respond
or replied 'don't know' to the question relating to local competitiveness, 21 per cent did so on the British competitiveness question and
29 per cent in relation to foreign competition. Similarly in the case

of technical efficiency: 15 per cent in relation to local firms, 20
per cent in relation to the rest of Britain and 20 per cent in
relation to foreign producers did not respond or replied 'don't know'.

Use of computers

Just less than half of the firms use a computer for some part of their
busines: 39 firms (45 per cent) use a computer for office administration purposes; 27 per cent (23) use a computer in the manufacturing
process; and 14 per cent (12) use a computer in the design process.
This pattern in the use of systems would suggest that a computer
assisted administration system is likely to provide the initial
introduction to the use of computer aided technology. Respondent
information proves to be sketchy in terms of the computerized systems
currently in use among these firms, but Stoll, Lectra, IBM, ICS Colour
Matching System, Universal, Barudan AMF and Equinox systems are all
mentioned. Pattern design, production and stock control, processing
orders, layout, colour matching, work ticket preparation, costing,
grading patterns and simple manufacturing jobs (e.g. attaching belt
loops) were some of the uses to which computers had been put.
   Of those firms which are non-computerized in particular areas, 13
have considered the use of computers for design, 15 have considered
the use of computers in manufacture and 27 have considered their use
for administration purposes.

Table 5.7
Use of computerized machinery by respondents

	No. of firms
Total	86
Of which:	
No computerized systems	44
Use some system(s):	42
Computer aided design (CAD) only	1
Computer aided manufacture (CAM) only	0
Computer assisted administration (CADMIN) only	18
CAD and CAM only	2
CAD and CADMIN only	0
CAM and CADMIN only	12
CAD, CAM and CADMIN	9
Total use of systems:	
CAD	12
CAM	23
CADMIN	39

   Just over a half, 51 per cent (27 firms), state that they would
seriously consider using a facility offering them low cost, easy
access to computerized facilities; 27 per cent (23) said they would

not use such a facility; 6 firms said they might and 13 firms did not respond. The responses indicated that 51 per cent (44) would use a training facility; 28 per cent (24) would not; 5 firms said they might; and 13 did not reply to the question.

Respondent information suggests that the most influential factor in deciding whether or not to computerize is firm size: 48 per cent (41) of firms chose the option 'firm too small for computerization to be cost effective'. The next most influential factor, 30 per cent (26), was the financial expense involved in computerization, and 23 per cent (20) felt that the time and money involved in staff training was inhibiting. About 15 per cent (13) of firms felt that they were inhibited by lack of knowledge of computer systems or their potential for the firm; 14 per cent (12) believed that there were no systems on the market suitable for their specific needs; 8 per cent (7) felt computers were too complicated to use easily; and 7 per cent (6) thought the introduction of computerization would be just too disruptive to the firm.

Table 5.8
Factors inhibiting computerization

Reason given	No. of firms
Firm too small	41 (48%)
Financial expense	26 (30%)
Time and money involved in training	20 (23%)
Lack of knowledge of systems	13 (15%)
No system suitable	12 (14%)
Too complicated	7 (8%)
Disruptive to the firm	6 (7%)

Approximately half of the respondent firms use a computer for some part of their business, but a large proportion of these firms have limited computerization in their office administration activities. About half of non-computerized, and slightly fewer than half of partially computerized firms at some time have seriously considered some form of computerization or further computerization. A number of suggestions as to how this could assist the design or production pro-cesses were made. Design was mentioned by five firms and manufacture by six. Administration was mentioned by four firms as was cutting. CADCAM, stock control and knitting were mentioned by two. Other suggestions were: distribution, production control, overlocking and pocket welting, having a fully integrated system, making up, pattern making, dyeing and finishing, packaging and layout design. One respondent wanted a computerized hammer for his continually misbehaving mainframe.

Factors influencing adoption of new technological systems

The decision to adopt computerized systems for administration, design and/or manufacture is assumed to rest on the assumption of profit maximization. The decision to adopt a particular system is similar to all investment decisions relating to physical and human capital investments. Additional problems may arise in the case of investment in new technological systems:

(a) frequently, firms will need to invest in both physical and human capital simultaneously, since their work force will be unfamiliar with the new systems;

(b) new technological systems may require fundamental changes in the working practices of the firm and in the structure of the internal labour market;

(c) new technological systems are usually introduced into the firm on a piecemeal basis - only rarely can firms afford to have a complete revamp of their capital stock. Integration of two or more generations of technology may lead to problems - for example, industrial relations and wage bargaining frictions between those who operate the new and those who operate the old systems, and bottlenecks where output from new machinery is produced much more rapidly than previously;

(d) the costs and benefits of using a new system will be more difficult to identify and evaluate;

(e) the benefits may only accrue over a much longer time horizon than firms normally accept for investments;

(f) the risks of investing in physical capital may be greater than buying new machinery using familiar technology, and the new technology may not live up to its expectations;

(g) financial institutions may be less willing to lend money to firms for investment where new technology is involved because of the difficulties and risks associated with identifying the benefits of the investment;

(h) the risks of investing in human capital to acquire skills in the use of new technology may also be greater - trainees may be more likely to fail to complete the training and they may be more likely to be poached by other firms.

Firms will adopt systems if, over some time period, the expected discounted benefits exceed the costs or where the expected rate of return to the investment satisfies the firm's minimum criterion. New technological systems will offer the potential to:

(a) reduce costs of production by reducing rejected outputs;

(b) speed up the output with a given unit of labour;

(c) increase the quality (and therefore value) of the output;

or achieve any combination of these.

There are a number of factors which might be expected to influence the decision to adopt new systems:-

(a) <u>Firm size</u> - the firm will need to be sufficiently large to benefit
    from the size of production runs which will make investment worth-
    while. Due to frequent changes in fashion, production runs in
    this industry are smaller than in other industries. A minimum
    size of 60 employees has been suggested as being necessary to make
    computerized systems worthwhile (Teague, 1987), though this is
    probably more applicable to manufacturing systems; small firms
    specializing in design, for example, could take advantage of
    design facilities using a micro-computer with a relatively small
    outlay.

(b) <u>Current trading position and expectations</u> - the firm may be more
    likely to invest in new technology if the firm's position in the
    market and its prospects appear good. Some writers have suggested
    the reverse: that firms will be forced to adopt new systems when
    traditional methods of production prove unprofitable.

(c) <u>Ownership characteristics</u> - it might be expected that firms who
    have adopted the status of limited liability will be more likely
    to invest in new technology than sole traders and partnerships
    because of the riskiness of the venture.

(d) <u>International competition</u> - it has been hypothesized that firms
    who compete in international markets are more likely to use best-
    practice technology, since they are prone to more vigorous
    competition.

(e) <u>Area of activity</u> - additionally, in the case of the textile and
    clothing industry, it might be expected that those firms engaged
    in knitwear production would be more likely to have adopted the
    new computerized systems than firms in other sectors of the
    industry since there has been more development of equipment and
    software in this sector.

(f) <u>Buyer-seller linkages</u> - the in-depth interviews with some of the
    firms revealed that the chainstores, especially one particular
    store, had been instrumental in encouraging the introduction of
    the new systems in order to control the quality of output and the
    delivery time lag, and to ensure accurate colour matching across a
    range of fabrics and over a long time horizon.

An empirical model

The model used to test the effect of these factors was a linear
probability model of the form:

$$Y_i = a + b X_i + e_i$$

where $Y_i = 1$ for adopters of new technology and $Y_i = 0$ for non-
adopters; X is a vector of individual attributes believed to influence
the adoption process; and e_i = an independently distributed random
variable with zero mean.

If we take the expected value of each dependent variable observation
Y_i

$$E(Y_i) = a + b X_i$$

and since Yi can only be either 0 or 1, then

$$Pi = \text{Probability } (Yi=1)$$

and

$$1-Pi = \text{Probability } (Yi=0)$$

Therefore

$$E(Yi) = 1(Pi) + 0(1-Pi) = Pi$$

The ordinary least squares regression equation is therefore interpreted as describing the probability that a firm will adopt new technology, given information about its individual attributes. An ordinary least squares regression estimation procedure does not constrain the probability estimates to lie within the range 0 and 1, although estimates outside this range are obviously inappropriate. Nevertheless, in these initial estimates outlined in Table 5.8 below, OLS results are reported.

Table 5.9
Factors influencing the adoption of CAD and/or CAM

Variables	CAD	CAM	CAD or CAM	CADMIN
Constant	0.03	0.23	0.27	0.16
Textile(1)	-0.22 (-2.38)	-0.29 (-2.72)	-0.32 (-2.89)	
Hosiery(1)		-0.22 (-2.11)	-0.24 (-2.25)	
Clothing(1)	-0.25 (-3.20)	-0.16 (-1.77)	-0.19 (-2.00)	
Size(2)	0.39 (4.11)	0.53 (4.45)	0.54 (4.38)	0.51 (3.90)
Chainstore(3)		0.24 (2.74)	0.22 (2.41)	0.23 (2.20)
R^2	0.25	0.37	0.35	0.32
n=81				

Notes: All variables (dependent and independent) entered as 0 or 1; t statistics in brackets;

(1) compared with knitwear firms;

(2) firms employing over 100 employees compared with those employing fewer;

(3) firms who supply a chainstore compared with those that do not.

The CAD equation is less reliable than the others since there are only 12 CAD adopters. The base firm has fewer than 100 employees, is a knitwear firm, and does not supply a chain store. The intercept of 0.03 suggests that the base firm has only a 3 per cent probability of having adopted CAD. Textile firms and clothing firms are less likely (22 per cent and 25 per cent respectively) to have adopted CAD systems than knitwear firms. The larger firms (employing over a 100 employees) have an increase of 39 per cent in their probability of having adopted CAD.

In terms of CAM, textile, hosiery and clothing firms are all less likely to have adopted CAM than the knitwear firms (29 per cent, 22 per cent and 16 per cent respectively), although the estimate for clothing firms is only significant at the 90 per cent confidence level. Again size has a significant effect. Firms employing over 100 employees have a 53 per cent increase in the probability that they will adopt CAM than smaller firms. Firms that supply chainstores have a 24 per cent increase in their probability of adopting CAM than those that do not.

If we look at adoption of new technology in either CAD or CAM, not surprisingly we find that again textile, hosiery and clothing firms are less likely to have adopted than knitwear firms (32 per cent, 24 per cent and 19 per cent respectively). Once again size has a significant effect with a 54 per cent increase in the probability of adoption for the larger firms. Chainstore suppliers have a 22 per cent increased probability of having adopted a computerized system for design or manufacture.

In terms of computer aided administrative systems (CADMIN) only the size of the firm and whether the firm supplies a chainstore have a significant effect. Firms with more than 100 employees are 51 per cent more likely to have CADMIN than smaller firms and firms that supply a chainstore are 23 per cent more likely to have such a system. There are no significant differences between knitwear, clothing, textile and hosiery firms.

It is perhaps also interesting to examine the factors which were found to have no significant effect on computerization:

(a) <u>Race</u> – a significant proportion of firms in Leicester are Asian owned. We found no significant differences between the white owned and the Asian owned firms in the adoption of computer aided systems.

(b) <u>Trade and prospects</u> – while it can be postulated that firms will implement new technological systems when the trading environment is favourable or when expectations for the future are good (or the reverse), we were not able to discern any significant difference between firms where turnover had increased in the last year and those in which turnover had fallen or remained static. Nor were there any differences between those that anticipated improvements in the next year and those that did not. However, firms with new systems may have introduced them in past years when their trading position was different.

(c) <u>International traders</u> – it was argued that firms which operated in international trading environments might be expected to have adopted computerized systems because of the more competitive conditions under which they operate. We found no evidence to support this hypothesis.

(d) <u>Difficulty with raising finance</u> - it might have been expected that firms which had experienced difficulty in raising finance would have been less likely to be able to invest in new systems. In practice this variable was found to have an insignificant (and wrong signed) effect. The positive effect might have been explained by the fact that investors in new technology might have been more likely to have sought and therefore discovered difficulty raising finance.

(e) <u>Owner managed firms</u> - there were no significant differences between firms which were owner managed and those which were not.

(f) <u>Age of the firm</u> - the age of the firm had no significant effect on the probability of adopting new systems.

(g) <u>Ownership characteristics</u> - there were no significant differences between different types of ownership - those companies with limited liability were no more likely to have adopted new technology than those without it.

The potential use of shared facilities

Around half of the firms in the survey said they would consider using a shared facility which offered the use/demonstration/advice of a computerized system. A similar proportion said they would consider using a related training facility. This is by no means an over-whelming response. A similar facility in Birmingham has proved successful and their use in engineering has been widespread, again with much success.

Summary of findings

A large proportion of textile and clothing firms in the Leicester City area are young, or relatively young companies which have adopted the legal status of the private limited company. A very high proportion of firms are single, independent manufacturing units, managed by the owner. Approximately two-thirds of owners are of a white ethnic origin and one-third of an Asian ethnic origin. An established group of Afro-Caribbean owners was not found operating in the textile and clothing manufacturing sector of industry within the Leicester City area.

A large percentage of textile and clothing manufacturing firms in the Leicester City area are small, or relatively small businesses in terms of staff employed. Approximately two-thirds of firms employ outworkers, at least from time to time, and approximately half of the firms employ freelance specialists, particularly for their design processes. The industry is dominated by female workers. Approximately one-third of firms are Asian owned and two-thirds are white owned with Asian owned firms employing a majority of Asian workers while white owned firms employ a majority of white workers, though most firms employ both white and Asian workers.

Textile and clothing firms in the Leicester City area depend heavily on banking institutions in order to raise capital. A surprisingly high proportion of firms have experienced no major problems when seeking to raise capital but, for the minority of firms that do, these problems have been chiefly associated with bank policy and practice. There is some evidence to support a cautious optimism for the stability and possible growth of the textile and clothing sector of industry in Leicester, with respondent information suggesting that more than two-thirds of firms are anticipating an increase in their volume of trade during the current trading year.

The survey has drawn responses from all four main sections of the textile and clothing sector of industry in the Leicester City area, as well as providing additional information about firms operating in industries which support and service those major sections. While many firms tend to engage in more than one productive activity, a high proportion of firms are involved in cutting and sewing, with similar proportions using their own fabric or buying in the fabric which they use to manufacture their products. There are fewer firms who produce their own knitted fabric to sell on, or who produce hosiery or body blanks. Far fewer firms produce fully fashioned garments. There is an extensive spread of main product lines manufactured by the firms in the Leicester City area. Respondent information indicates that approximately one-half of Leicester firms subcontract work out to other companies.

Textile and clothing firms in the Leicester City area use a variety of different outlets to sell their products, chief among these being wholesalers, large chain stores, single retail outlets and mail order companies. These outlets are situated throughout Britain, but a sizeable proportion of firms sell some of their products locally and approximately one-half of firms are active in the export trade. Exporting firms trade in a variety of different countries ranging from the EC and Scandinavia to North America, as well as to Africa and emerging nations around the world. The largest single market for export goods for the firms is Eire, and a large proportion of total exports go to EC countries generally.

Firms based in the Leicester City area use a variety of sources to obtain supplies of yarn and fabric, but a high proportion of these supplies come from locally-based companies. This obviously offers the opportunity to influence the purchasing patterns of local suppliers of yarns and fabrics.

A high proportion of firms in the survey either rate their efficiency poorly or are unable to accurately compare their firm's competitiveness and technical efficiency with competitors in other parts of Britain and abroad, probably due to a lack of knowledge about such competitor firms.

The most significant factors influencing the adoption of computer assisted design, manufacture and administration in the Leicester clothing and textile industry were the size of the firm and whether or not the firms supplied chainstores. This does not suggest that small firms are innovative but rather the reverse. It also implies that chainstores exert a significant discipline on their supplying companies which can perhaps be compared with similar processes which are influencing the components suppliers to the car assemblers. The knitwear firms were much more likely to adopt computer assisted systems for manufacture than clothing, hosiery or textile companies and also more likely to adopt computer assisted design systems than textile and clothing firms.

There is some evidence to suggest that a high proportion of firms would seriously consider using a shared technology facility if one were established in the area, with low cost and easy access coupled with employee training. Few firms offered suggestions as to what particular processes they felt would benefit from computerization, but production control and general office administration featured prominently among those suggestions which were received.

6 The case study findings: Leicester textiles and clothing firms

Introduction

The purpose of this phase of the research programme was to supplement
and amplify the information obtained from the general survey by
focusing on specific aspects of the structure and organization of the
textile and clothing manufacturing industry in the city of
Leicester. From a close inspection of the experience, knowledge, and
opinion of people working directly in the industry in Leicester, it
was hoped that general operating trends might be identified, and
perceived problems and constraints explored. A careful selection of
case study companies was made from among those firms that had
responded to the initial survey. The criteria for selection was
three-fold. First, selected companies were required to represent a
cross section of the industry in Leicester in respect of type of firm
and the technology used. Second, a slight emphasis was to be made in
favour of the smaller company, since these companies were believed to
be more likely to need assistance and third, companies were to be
sought whose initial survey responses indicated information of special
interest to the research programme. A careful re-selection was made
in four cases of cooperation refusal.
Twelve companies were finally selected for interview. Of these
companies, the majority are single, independent units of private
limited companies, managed by their owners. The ages of the companies
range from an establishment year of 1903, through to young companies
founded in the 1980s. Eight of the companies have a white ethnic
management, and four have an Asian ethnic ownership and management.
Firm size ranges from the very small through to the medium-large in
terms of numbers of employees, with an emphasis on those firms
employing fewer than 100 people. Four of the companies employ fewer
than 20 people; three firms employ between 20 and 50 people; three
employ between 50 and 100 people; one firm has a labour force of 300,
and another employs over 600 people.

All four broad categories of firm (clothing, knitwear, hosiery and textiles) are represented in the sample. Four companies are solely engaged in the clothing sector of the industry; three companies solely in the knitwear industry; two in the hosiery trade only, and one company manufactures fleece fabrics in the textile sector. The remaining two companies (the two largest case study companies) are each involved in more than one sector of the industry. One firm manufactures for the clothing, knitwear and hosiery trades, and the other for all four sectors of the industry. The methods of production adopted by the firms include all those featured in the survey questionnaire, except that of 'part finished garments'. All the main product types, and some specialist ones, are represented in the manufacturing output of the companies.

Three of the case study companies use computerized manufacturing and office administration technology, and one of these companies also has access to computerized design facilities. Four further companies use computers to help with their administration and management functions but do not use computers in their manufacturing or design processes. The remaining four companies are totally non-computerized at design, manufacture and administration levels.

The results of the in-depth investigation of the twelve companies are given in the following six sections, each section indicating the general trends which have been identified from the evidence.

Interviewee and company history

Family links and coincidental first-job experience would seem to be the two most powerful factors influencing personal involvement in the textile and clothing industry. Six of the twelve interviewees joined existing family businesses, mainly headed by their fathers, and now hold senior management positions in their respective firms. All four Asian interviewees are represented in this group, as are all four sectors of the industry. Four interviewees began their involvement in the industry on leaving school, two to take up traditional apprentice-ships, one in knitting skills and one in tailoring. The remaining two interviewees, both women and both in managerial positions in the clothing sector, came into the industry by chance, by answering advertisements for administrative and managerial posts.

Respondent information revealed a variety of circumstances occasion-ing company start-up. The particular circumstances surrounding the birth of the two longest established case study companies, both clothing firms, are now lost in history, but for the other ten companies, two recurring sets of circumstances were revealed.

The first of these revolves around interviewee desires to use individual skills to enhance personal advancement and autonomy. In this respect, three of the ten respondents reveal the acknowledgement of a valuable skills base which they wished directly to exploit, both in terms of their own best interests and in terms of the interests of their respective sectors of the industry. Six of the ten respondents (representing all four sectors of the industry) report that they experienced a conscious desire to 'be my own boss and control my own life', and that this need was a principal factor driving them on into company ownership (see Bosworth and Jacobs, 1987). All six report that, despite the difficulties which they encounter in their working lives, they would not, today, wish for a change in this personally autonomous position. Three interviewees report that it was the

combination of both these desires, that is to utilize acquired skills
and to become occupationally autonomous, which formed the basis for
formation of their own companies.

The second of these two start-up themes emanates from the position
in which respondents found themselves during one or other of the
recessions experienced, especially by the knitwear sector. Redundancy
in times of recession, and the parallel need to 'earn a decent
living', was reportedly the main spur which prompted four of the
interviewees to start their own companies, three of these in the
knitwear and one in the clothing sector of the industry.

As something of a side issue, but nevertheless of interest to the
main concerns of the case study investigation, respondent information
was searched for a correlation between entrepreneurial ambition and
company success. While a statistical analysis on such a small sample
cannot be held to be predictive of a general trend within the
population, no significant correlation could be found between
company success, in terms of actual and predicted volume of trade,
and a personal 'own boss' ethos and ambition.

Operating problems and advantages

Interviewees were unable, immediately, to offer many 'pros' to working
in the industry. However, three respondents felt that their
companies, having managed to corner a specialist market, had no need
to 'chase' orders or find new customers. Reportedly, their main
problem centred on meeting an ever increasing demand from existing
customers and coping with, or turning away, additional orders from
new ones. Two respondents, both manufacturing knitwear, felt that
involvement in a swiftly changing fashion industry made for a
challenging and interesting career, and another respondent felt that
it was advantageous operating in the nil VAT rated children's clothing
trade. Three respondents (all operating in the clothing sector) felt
that there were no 'pros' to working in the industry.

Two-thirds of all case study respondents reported that a
professional approach to financial control, marketing, and forward
planning is crucial in maintaining company success and growth, and
gives their company a competitive edge over other businesses in the
Leicester-based industry. Maintenance of tight quality control of
specifications, especially in respect of colour specification, was
considered to be important by seven respondents in keeping them ahead
of the competition. Six respondents emphasized the need for companies
to respond quickly to changes in demand and for a personal service to
be offered to customers. The utilization of specialist skills
reportedly kept four companies ahead in the clothing, textile and
knitwear sectors of the industry, while a textile and a clothing
company stated that a constant updating of machinery and plant is a
crucial element of company success.

Perceived problems were many and varied. There were four major
issues identified, the most commonly mentioned being that of fierce
competition. Eleven separate responses were received from respondents
in this respect (from all four sectors of the industry), five
emphasizing the difficulties of competing with cheap imports from
Third World countries and from the poorer countries in the EC. Seven
responses emphasized the highly competitive nature of the home
trade.

A secondary but substantial problem, particularly for manufacturers operating in the clothing sector of the industry, is reportedly that of staff recruitment, both skilled and semi-skilled. Respondents also report a general unavailability of committed people willing to work in an industry which has a negative image and which is unable to offer status or 'perks'.

New fashion demands for a wider variety of colour specific materials, both fabric and yarn, and large-retailer requirements for colour-ratio manufacture and packaging are a growing problem particularly for the knitwear and clothing companies. Those companies involved in the manufacture of knitwear and hosiery also report difficulties associated with the use of contracted dyeing companies, and the acquisition of yarn of consistent quality and specified colour shading.

There is some evidence to show that companies involved in all four sectors of the industry are experiencing problems in acquiring basic materials of a sufficiently high quality in the quantities required. Evidence also exists to suggest that companies are increasingly unwilling to hold large stocks of high quality materials as these stocks qualify for taxation. Stock building to avoid shortage of supplies now has a prohibitive cost.

Over half of the case study companies, in all four sectors of the industry, reported that it was not at all difficult for them to keep themselves well informed about competitor activity and performance. In what is in reality a small community, the grapevine is very active, easy to access, and believed to be reasonably accurate. Three respondents from the knitwear and hosiery sectors, however, felt that because they had little contact with other people in the industry it was difficult for them to assess accurately their performance against that of others. One knitwear firm which specializes in design and technical services consultation to the Third World, reported that with so few companies involved in this area of work it was difficult to make direct comparisons of performance and that they rely on product sales for a guide to company effectiveness.

The labour force in the industry

Evidence shows that the companies recruit their labour force from the immediate locality, using a combination of strategies for making contact with appropriate potential staff. The local press, Job Centres and the company grapevines are all used extensively. Two companies which have recently recruited for management positions reported a more geographically extensive trawling for appropriate candidates.

Two-thirds of case study respondents reported that they have recurrent problems in the recruitment of staff and that the most difficult person to find is the conscientious worker with good quality traditional skills, who is willing to work and is capable of working to exact specifications without needing much supervision. Individuals with the ability to take decisions and to detect faults are also reported to be difficult to locate. Case study evidence shows no specific differences in labour recruitment between firms in the four different sectors of the industry but it is interesting to note that five of the twelve case study respondents emphasize a perceived shortage of what might be termed 'positive characteristics' in potential and existing staff members. Characteristics such as reliability, responsibility, willingness to work and conscientiousness were frequently mentioned by these five respondents who clearly regard such aptitudes as 'skills' which they require their work force to possess. It is evident, therefore, that the definition of skill used by these firms is extremely broad, encapturing attitudinal and personal qualities in addition to the ability to undertake specific tasks.

The results reveal a division between those respondents who consider that staff trained in the traditional skills qualify as 'skilled workers', and those who do not. Two-thirds of the case study sample felt that workers with traditional skills should be categorized as 'skilled' workers. One-third of respondents felt that their labour force should be categorized as semi-skilled. This latter group of respondents reported that it is the machines that do the work; all that the people have to do is to operate them properly and this does not qualify as skilled work. There is no case study evidence suggesting the existence of particular differences in the levels of the work being done by staff in either group of respondents. Information received from interviewees whose companies use computerized manufacturing technology somewhat surprisingly came out three to one in favour of categorizing workers with traditional skills as being 'skilled' workers, although there are some predictions that the high technology machines are leading to an increasing de-skilling of the work force. Those respondents who defined their work force as being semi-skilled also emphasized the need for aptitudes of reliability and conscientiousness in their staff. These respondents operate in the hosiery and clothing sectors of the industry.

The case study information provides evidence of the deep rooted and continuing widespread existence of an informal and potentially inexact procedure in use for the training of company staff using non-computerized technology. Nine of the twelve case study companies reported that the method of training undertaken is of the 'sit by Nellie' type where new recruits sit next to an experienced worker, initially to observe how the work should be undertaken, and later so that the experienced worker ('Nellie') can oversee the newcomer's work. Eight respondents emphasize that 'trainers' are carefully selected from experienced personnel, and new staff are taught basic skills and company procedures by these selected people. Two respondents reported that they only recruit suitably trained people and that new staff are instructed as to procedural needs by existing and dependable personnel. Only one company, a specialized clothing firm, reported that it recruits only professionally trained staff whose basic skills are then further developed by the company directors and by very experienced staff.

For those four companies employing the new technology, however, this situation is reversed. Staff training in respect of computerized manufacturing machinery in both the clothing and knitwear sectors, is selectively given by computer manufacturers' training experts, although staff using non-computerized machinery continue to be trained by colleague operatives. In the same way, those companies using computerized office administration systems rely, in all cases, on manufacturer training however inadequate this may turn out to be.

Only three out of the twelve case study companies reported the existence of graduates on the company staff. These three companies employ business studies graduates in their management team and one company also employs a graduate textile designer.

The technology employed

Respondent information reveals a wide range of traditional manufacturing machinery being used by the case study companies including flatbed and circular knitting machines, high-speed sewing machines, and bandsaws and knives for cutting.

The case study evidence points to a high use of foreign-made machinery among firms in the Leicester industry. Clothing sector firms appear to favour Rimoldi, Singer, Union Special and Brother sewing machines, and the knitwear case study companies are using Conti, Union Special, Wildt, Rosso and JDR Dubied circular and flatbed machines. The case study textile firm depends entirely on foreign-made machinery, using the Orizio two-thread and the Mayer three-thread fleece machines from Italy and Germany respectively. However, three of the knitwear case study companies are using some British-made Bentley flatbed and circular knitting machines. One of the case study firms, which specializes in the design of simple knitwear technology for Third World use, emphasizes the need for reliable manual and electrical machinery which requires minimum maintenance and which supports the continuance of the traditional skills. (This firm claims to have developed a 'no needle' machine which is not in operation or available for inspection because of lack of patent protection). Evidence shows that the most common criterion for choice of make of traditional types of machinery is reliability, closely followed by speed, ease of maintenance, and availability in terms of supply of models, parts and servicing.

Responses suggest that many of these manufacturers are aware of the possible or potential implications arising from non-adaptation to computerized technology and are reassessing their position in this respect. The main benefit of the adoption of computerized technology for all four sectors of the industry is seen by the case study companies as being the opening up of new market possibilities, followed by the production of a better quality garment. The reduction of waste and the saving of labour are emphasized by clothing and hosiery sector manufacturers. Only two of the non-computerized case study companies identified the expansion of design potential as being a primary benefit of adaptation to computerized technology. (In view of other respondent information this is not such a surprising finding. This point is referred to again in the next section). The most commonly expressed inhibitors to computerized adaptation remain cost effectiveness, especially for the small company, and loss of company autonomy, especially in terms of overdependence on 'outside' expertise should the computerized equipment break down.

The technology currently in use among the four computerized case study companies includes: Schima glove making machines; IBM System 36 and Union Special lay-planners; San Giacomo sock making circular knitting machines; Universal MC 640 flatbed knitting machines; and Bierrebi cutting and Godrich welting machines. All four computerized companies report that they adapted to high technology for reasons of speed and flexibility especially in respect of design change and correction of design fault, and the expansion of their market base was also an objective for three of them.

While these computerized companies consider that high technology adaptation was costly, their information suggests that it is felt to have been worthwhile in terms of initial objectives and in terms of staff approval and reliability. The demerits would seem to centre around the ongoing capital and running costs associated with its

acquisition and the hidden costs involved with the inability of the
high technology equipment to cope with substandard materials. High
maintenance costs or repeated computer breakdown do not feature as
being problematical experiences for these firms.

   Among the seven companies that use a computer to help with administrative and financial systems and processes, by far the biggest
problem would appear to be the inadequacy of computer manufacturer
back up services, especially those relating to the software systems in
use. Two respondents refer to the generally poor level of software
servicing obtainable despite assurances to the contrary at the
purchase point. Two other respondents report that the software which
they were persuaded to purchase turned out to be inadequate for their
needs, and one further respondent reports that the software purchased
actually produced inaccurate and misleading data due not to operator
or company error but to the undeveloped nature of the software
itself. Two respondents refer to the expensive running costs of the
computerized systems, a factor apparently not mentioned at the time of
purchase. Three companies claim to have experienced incapacitating
and recurring breakdown in respect of their office administration
computerized systems, and one respondent emphasizes her repeated
experience of a markedly intolerant response to requests for help from
the software company's advisory service.

   Given that case study information shows that these seven companies
are using a wide variety of different computerized office administration systems, there would seem to be a widespread need for an improved
response from computer manufacturers in respect of both purchase
advice and quality of after sales servicing. Case study interviewees
are expressing a clear need for a comprehensive, neutral, and expert
advice service for initial choice of system, for operational problem
solving, and for staff training for those companies that are either
contemplating adoption of office administration computerization, or
are planning an expansion of current computer use.

   However, despite recurring faults and poor servicing and training,
respondents report that, generally speaking, their computers do meet
speed, time saving, data availability and labour saving objectives by
accommodating the accounting, invoicing, budgeting, costing, and wage
payments of the companies concerned.

Manufacturer/customer relationship

The case study information shows that the twelve companies produce a
wide range of products for a variety of different markets ranging
from market traders and the 'bottom end of the market' through to
nationally known high street chain stores, high-fashion boutiques and
West End quality fashion houses. The products manufactured by the
case study firms encompass all forms of under and outerwear, both in
knitwear and woven fabric, for men, women and children; specialist
sporting and protective riding wear for all sections of the market;
informal sportswear and leisurewear for men, women and children;
traditional and high-fashion hosiery, underwear and nightwear; knitted
fleeces and fabrics, and gloves.

   Case study companies, in all four sectors of the industry, produce
for customers that include the well known names of the high street
retail chain stores including the more exclusive 'West End' stores in
a few cases. The hosiery and, in particular, some clothing firms
rely to some extent on sales to the major mail order companies

including some well known specialist companies with good reputations for quality. A specialist clothing company which makes protective riding wear, sells to specialist riding retailers countrywide and to the Jockey Association, the Police Force, the Royal Mews, the Household Cavalry and to various film companies. On a 'high' note, this company also claims an 'HRH' among its customers. Four of the companies, one operating in each of the four sectors of the industry, report that the main bulk of their production goes to small retailers and market-stall traders at the 'bottom end' of the market.

All twelve case study companies report a variety of customer imposed conditions and standards relating to their manufacturing processes and production techniques. In effect, respondent information suggests the existence of a large-retailer control of the manufacturing units involved in the knitwear and clothing sectors and, to a lesser extent, also in the hosiery and textile sectors of the industry. Such control would seem to extend potentially to almost every stage and aspect of the manufacturing process including the choice of sources of supplies; the design, sizing, quality and colour of the products themselves; labelling, packaging, storage, take up and delivery of orders; and in some cases to the administrative systems adopted. Some respondents report that many imposed customer requirements work directly against the manufacturers' best interests but that failure to comply inevitably results in the loss of orders.

In this respect there are two types of policy adopted by the large buyers. These two policies are referred to as 'hands on' and 'hands off' approaches. The extreme form of 'hands on' approach involves the buying firm having a close involvement in the working conditions, pay, technology and decision making of their supplying firms. The extreme form of 'hands off' approach involves the buyer taking no interest in how the goods are produced but making orders on the basis of the style and quality of the goods, the price and delivery time.

While some of the case study companies report that they welcome orders from the large retail chains however restricting or difficult the imposed conditions may be, five respondents specifically indicate that their company has made a policy decision not to trade at the top end of this market. For these firms, one large retail chain in particular has been abandoned as even a potential trading partner because of the hold which this company gains over its supplying manufacturers. The reasons for the decision relate to the reduction in autonomy that results from accepting the conditions of the retail chain, the disruption to orders for other companies because of the priority they are required to give that firm, and the resultant dependence on that firm for orders as other customers go elsewhere when their orders are consequently given a lower priority.

The most frequently imposed conditions relate to the maintenance of quality control in terms of materials, design, making up, colour matching, and labelling and packaging. All the case study companies report that their customers expect most, if not all, of these aspects of product manufacture to be carried through precisely to specifica-tion. Labelling and packaging specifications are not so prevalent at the bottom end of the market but are becoming an increasing source of difficulty to firms producing for the large retail chains. At this end of the market, retail design specification is now so exact and organized that the hosiery, clothing and knitwear firms supplying the high street retail chains no longer tend to employ an 'in company' designer, the role of design having shifted to the chain stores.

Respondents say they have no need for, or outlet for, company designed products in these days of fast changing and precise fashion requirements of the retail trade.

Nine of the twelve case study companies report that they have changed, or adapted existing company manufacturing or administrative procedures, in order to meet the specific requirements of their large-retailer customers. One clothing and knitwear firm which manufactures for arguably the largest and best known high street retail chain, were required to buy in expensive and specialist lighting equipment for colour matching purposes. The choice of purchase of the equipment was specified by the retailer who also supervised installation and the training of operatives. Three case study companies in the clothing and knitwear sectors have been required to respond to customer expectations for the manufacture and packaging of products and mixed colour ratios. These companies report that this takes more time, more space, and that in effect they are organizing the retailer's counter for them.

Five companies in hosiery, clothing, and knitwear have been obliged to accommodate customer imposed design changes and exact specifications - a development which had implications for the employment of designers. Two respondents report that their companies made manufacturing changes in respect of retailer requirements for oversize clothing. These two companies, one hosiery and one knitwear, explain that manufacturing extra-large garments has implications for the full use of standard fabric widths and for the maintenance of competitive prices. Three companies, all in the clothing sector, have been required to change their administrative processes in order to meet customers' computerized invoicing and product code requirements.

One clothing company has been required to adapt to a palletized packaging and delivery system to meet imposed customer demands, and the case study textile company reports that, contrary to its own standards, it has increasingly been forced to manufacture fleece fabric of a poor quality. This is not to this company's advantage in that a poor quality fleece takes longer to knit and is therefore more expensive to produce but sells at a lower price than fabric of a higher quality. One clothing company reports that one of its biggest problems is that of accommodation to customer take-up options on ordered goods. Its large-retailer customers increasingly demand the facility of accepting or rejecting a part or even the whole of an order to meet their changing market circumstances.

Possibilities for local authority involvement and shared facilities

Interview experience revealed that case study respondents were generally loath to become involved in a detailed examination and assessment of company involvement in a proposed shared high technology facility similar to that set up in Birmingham by West Midlands County Council. Similar shared facilities have been introduced in other areas in other industries, particularly in CAD facilities for engineering firms (see Bosworth, Jacobs and Lewis, 1987 and Section 7 below).

On being persuaded to assess the potential effectiveness of a nominated variety of different services which might be incorporated in such a facility, respondents opted primarily for four suggested features.

(i)   The most popular option, chosen by over two-thirds of
       respondents, was for 'hands on' experience of computerized
       equipment. Demonstrations were seen to be less helpful in that
       they could be confusing. It was the 'hands on' experience which
       was perceived to be most effective because of its immediacy.

(ii)   Two-thirds of the firms were in favour of options offering
       information and advice on available technology and company-
       specific advice on choice of appropriate technology. This
       option carried the proviso that such advice and information
       should be neutral, expert, and given by persons with a
       background and additional expertise in the manufacturing base of
       the industry. All such advice should be of a company-specific
       nature. General advice was seen by respondents to be useless at
       this level, and non-neutral advice was seen to be already fully
       available in the form of computer company sales pitches, most of
       which proved, with hindsight, to be misleading.

(iii) Half the respondents opted for a short term rental service for
       use on own company premises, this proving more popular than 'by
       appointment' use on facility premises, or short term expert job
       supervision of company manufacturing processes.

(iv)   Individual suggestions for additional facility options were rare
       but did include the need for an expert follow-up maintenance
       service on the new technology and a computerized design
       facility, centrally situated, for use especially by knitwear
       firms. Such a design facility would need to be compatible with
       existing computerized manufacturing technology currently
       operative within the city's industry base.

In terms of technological aid the majority of respondents felt that
the local authority could do very little to help them directly in
adapting to high technology. Eight out of the twelve case study
respondents indicated that they did not feel that this lay within the
remit of the local authority and that the council had other equally if
not more important responsibilities to fulfil. However, one
respondent felt that the city council could make an effective contribution towards company awareness of this issue. This hosiery and
clothing firm representative suggested that the city council publish a
'new look' business magazine, one which would be interesting, easy to
read, colourful, and which could give up-to-date information on new
technology. This respondent remarked on the general dullness of
most business magazines currently available on the market and the
difficulty involved in understanding some of the language employed by
these journals by hard pressed, non-technical businessmen and women.
In terms of other ways in which the local authority might help the
city-based textile and clothing industry, an immediate and widely
varying response was forthcoming. Two-thirds of respondents, from all
four sectors of the industry, argued for city council involvement in a
bid to improve the image of the industry in general both at home and
abroad. The argument was made that raising the image of the industry
might increase the status of the associated activities and attract
much needed talented and committed young people into the sector.
Three respondents argued that promotion abroad, not only of the
industry but of the city and its facilities and services in general,
would be a positive way of attracting foreign buyers' new business

which small companies are themselves unable to attract through lack of resources. A combination of planning, editorial, publicity and advertizing expertise, supported by imaginative and modern resource equipment, could be coordinated by the local authority to the benefit of the Leicester-based industry. Such resources would be especially useful at trade fairs and exhibitions sited abroad. Small-business personnel especially were often unable to leave a business for long enough to visit a trade exhibition in order to make new contacts, attract new business, or orientate to new developments. Nor can they afford to produce up-market promotional brochures - perhaps in foreign languages. Respondents think that any help which the local authority felt able to give in this direction might make a very real difference to the operating potential of small businesses in the Leicester-based industry.

Two respondents additionally argued that help to combat gender differentiation and racial prejudice within the industry would improve the image of an industry seen by many to maintain entrenched attitudes and outdated systems of management. One knitwear firm respondent suggested that the city council could establish creches for the children of working women and one clothing firm representative suggested that the local authority could educate employers in good employment practices. These liberal and socially aware attitudes by these two respondents may not be typical of management views in the industry and Leicester.

Four respondents felt that the local authority could be of direct help in terms of training. One respondent from a clothing company felt that many small business managers lacked the requisite management skills, especially in terms of short and long term planning, marketing, and preparing proposals. It was accepted that high quality business management training was currently available from other sources but it was felt that, at the very least, there could be a local authority role in sensitively promoting management training courses to encourage small business take-up of the help available. Three respondents thought that the teaching of traditional skills in the clothing and knitwear sectors was an appropriate area for local authority involvement. While industrial training boards traditionally fulfil this purpose, respondents felt that the city council could adopt an initiating and coordinating role for the expansion of the training in such skills. A respondent representing a highly specialized clothing company felt that the city council should use its influence to revive apprenticeships in the traditional skills.

Three respondents suggested different means by which the local authority could improve the financial viability of firms in the industry. A clothing and hosiery company believed that the creation of a credit union, offering short term loans to overcome cash flow difficulties, would be particularly helpful to small companies. This same respondent also proposed the establishment of a finance cooperative organized by the city council, with built-in advice and control features. A textile company representative felt that the city council could make low interest loans available which would enable small companies to move through the often problematic four-to-five year expansion period. This respondent reported that, in his opinion, it was during this period that small companies were especially vulnerable because of a tendency to overtrade. The third respondent, active in the knitwear industry, felt that the city council could encourage Third World self-help by providing finance for schemes, organized in Britain, for operation in Third World countries.

One respondent, from the clothing sector, suggested that the city council set up a 'Grievance Bureau' for problem solving, using expertise already existing in the industry. In a similar vein, a textile sector representative emphasized the need for the local authority to establish firm channels for official proceedings, such as 'change of use' and 'permission for twenty-four hour use' applications. This respondent stated that it was his belief that at present these proceedings are open to corruption, a situation which can lead to feelings of resentment within certain parts of the industry.

Five respondents from all four sectors of the industry want the local authority to concern itself more with the provision of business premises. One knitwear company representative pointed out that in her capital intensive sector of the industry, good, large, centrally located business premises were at a premium. This particular company was prevented from expanding its business because they could not find appropriate premises to purchase. A textile company representative supported this view in general, and a clothing sector respondent stated that, in his labour intensive industry, business premises of an appropriate size and quality were difficult to find. One representative reported a monopoly of large business premises by a handful of operators who hire out the premises and undercut council rents. One respondent from the knitwear sector argued for the establishment of longer periods between rent reviews.

Other suggestions for local authority intervention included the support of sensible import regulations, although this clothing sector representative recognized that this was more a matter for central, rather than local government action. Two respondents urged the city council to take measures to ease car parking difficulties in inner city business locations, and one respondent argued that the most crucial broad-based role for the local authority was that of promoting the overall well-being of the whole community in Leicester, including the business community. In this respect, this respondent stated that he felt that the city council was active on behalf of the population.

In closing, and perhaps inevitably, exactly half of all respondents representing all four sectors of the industry, urged a reduction of the rates. Leicester is one of the lowest rated areas in the country and typically rates account for between one and two per cent of turnover for firms in Leicester.

7 Shared facilities and small firm behaviour: West Midlands case studies

Introduction

This part of the book focuses on the empirical results of case studies carried out as part of the ESRC funded project. The intention of this phase of the research programme was to focus on the actual experiences of those currently involved in established centres for shared high technology facilities and on the histories and perceived problems and successes of those centres. From information received from the chosen case study projects it was hoped that common experiences, trends and constraints could be identified, and some future applications for high technology systems and general and specific project developments could be indicated.

The research methodology

A four stage methodology was adopted in order to obtain the case study material. This consisted of firstly, a preliminary telephone search to elicit information on existing high technology shared facilities countrywide from which the case study projects would be selected. Secondly, the design of two questionnaires, one to gather information required from the selected case study projects and one to obtain details of user firm utilization of case study project facilities. Both questionnaires were piloted and amended as required. Thirdly, the selected projects were visited, viewed, and recorded on-site interviews were conducted with project staff and fourthly, a small random selection of user firms was surveyed by telephone.

A decision was taken to restrict the case study work to a limited geographical area of the country. Apart from time constraints, it was felt that the adoption of a procedure which focused on the facilities available in one region of the country would enable a meaningful identification of patterns of funding and planning involvement, and a

close comparison of existing services and perceived problems and
constraints which could be set within a common industrial and
employment context. The West Midlands Region was identified as
being a fertile area for investigation for three main reasons.
First, preliminary information revealed an intense clustering of
funding local authorities, large scale industry, and business and
educational establishments in the region. Second, there was evidence
of established local authority/educational establishment cooperation
in the inception and operation of shared high technology facilities in
the area. Third, there was initial evidence of a common rationale
for interest and investment in high tech industrial and business
conversion borne out of the need to combat the heavy losses in
traditional, male industrial occupations experienced in the area over
the past decade. Therefore, while some general information was
obtained on existing facilities throughout the country, and more
detailed information was elicited on the East Midlands area, the
emphasis of the case study evidence is centred on projects and user
firms based in the West Midlands region. It is fair to say that the
spatial emphasis might well have been different with hindsight, given
the subsequent funding of work in the Leicester area.

The case study questionnaire was designed to elicit specific
information, through personal interview, on: funding sources, funding
and staffing levels, the range of project equipment and services,
constraints on project expansion, and plans for future development.
Additionally it sought general information on project planning and
interorganizational cooperation, perceived operational success or
failure, and the range and composition of user organizations. Using
the questionnaire in face-to-face on-site interviews with project
staff afforded the research team the opportunity to view the high
technology systems themselves, often in operation, and also to obtain
a feel for the atmosphere engendered by the different projects.

Five high technology shared facilities were selected for in-depth
case study investigation. All are located in the West Midlands, and
all offer a wide range of training and consultative services including
'hands on' experience of high technology systems and machines. All
five projects have become established during the last six years. One
project is currently self financing; one is working towards this
position, and three continue to utilize funding sources involving a
mixture of local authorities, educational establishments, national
government agencies and user organizations. Three of the project
facilities are located within higher educational establishments;
another project has adopted the form of a limited company, and the
fifth is based within the national ITEC initiative. Four of the
projects offer services to a wide cross section of business and
industry, but mainly to the general and electrical engineering
sectors, and the fifth project offers services especially aimed at
local textile and clothing firms. The results of the in-depth
investigation of the five case study project facilities are given in
five sections, each section indicating the general trends which have
been identified from the evidence.

This is followed by a discussion of the results obtained from a
brief survey of user firms, undertaken in order to test whether the
perceived successes reported by the projects were confirmed by the
user firms. A secondary purpose was to gather limited information on
problems which user firms may have encountered during high technology
transition within the firm, and to examine the perceived importance of
training to the user firms.

The shared faciltiies

Project planning and establishment

Case study information would suggest that the shared high technology centres for business and industry are of relatively recent origin. All the case study projects have been planned and have become established during the past six years, with the exception of one which began a tentative development in 1979. This is borne out by information gained from the preliminary telephone search which located seven additional projects in the Midlands area which have become established within the past six years, one established a little longer, and five projects currently in the planning stage or about to come 'on line' in the near future.

There is evidence to suggest that the planning of shared high technology facilities is becoming more organized and coherent as awareness of the capabilities and applications of computerized systems increases. The research revealed the somewhat tenuous nature of the 'planning' and inception of these projects in the late seventies and early eighties. Three of the projects studied evolved gradually out of already existing educational and training practices, and the planners of the 'older' projects admit to having no clear conception of the potential and importance of computers to industry and business at the time that they began project development. There is some evidence to suggest that the eventual establishment and successful operation of the projects was to some extent dependent on the tenacity, enthusiasm and the 'hunches' of one or two key people in the existing organization. Three of these 'older' projects undertook no research during the planning phase, and respondents of all the case study projects report that their project was not modelled on any other existing shared facility. The 'youngest' of the case study projects, which became operational in the summer of 1986, did undertake systematic research in support of project planning, and information obtained from the telephone search would indicate the existence of a more concerted movement among current project planners to share information and research fully the potential for shared facilities in their localities.

There is strong evidence of a common motivation for the establishment and further development of shared facilities from the case study projects. The main objectives of all the projects studied are to help to combat growing unemployment in the traditional industries in the respective localities, and to provide 'better quality jobs' in industry and business. Project emphasis here would appear to be two-fold; one, to help make industry become more efficient and competitive and thus increase employment potential, and two, to upgrade worker skills both for those currently in work, and to enhance the employment prospects of those currently unemployed.

Project funding and operational costs

Evidence from both the case study projects and from the preliminary telephone search suggests a somewhat incohesive and haphazard structure and nature of shared project facility funding. The sources of funding for the case study projects are both varied and numerous but local authorities of various kinds would appear to be an important source of pump-priming funding, and local authorities, in conjunction with government agencies, the main source for ongoing operational

funding. Initial funding from industry or business would appear to be
low or non-existent, with the exception of one project which received
substantial funding from a national banking institution. Two of the
three projects operating within the auspices of educational establishments have the advantage of a 'hidden' source of funding in that
mainframe computing facilities are funded by the college, as are the
salaries of lecturers and technicians who contribute to the services
of the projects in a part time capacity. In more recent years, three
of the projects have acquired access to substantial funding from the
EEC and intimate they will be seeking increases in future funding
from this source. Respondent information would seem to indicate that
European Fund money might assume considerable importance to the
projects in future years.

Despite the variety of funding sources available to the projects and
the different circumstancs under which the projects operate, research
evidence nevertheless suggests a number of common trends among the
projects studied.

There is a paramount need for substantial pump-priming and ongoing
funding for expensive equipment and for technically experienced and
competent staffing. This is confirmed by information obtained from
the preliminary telephone search. All the case study projects
required pump-priming money varying from £15,000 for one project to
make a tentative beginning, to between £125,000 and £170,000 for the
other projects. The need for substantial funding is highlighted by
the fact that the project receiving £15,000 pump-priming money was
unable to develop its services until it received additional funding of
£100,000 three years after inception. Current annual running costs
vary between £125,000 and £785,000 dependent largely on the number of
staff employed and the nature of the equipment acquired by each
individual project.

All five case study projects report that while project establishment
was not hindered by imposed funding restrictions, all pump-priming
spending had to be in line with pre-specified project objectives. One
project experienced a funding restriction, or pre-condition, which
turned out to be highly efficacious to the project. The restriction
was that a set proportion of the funding which came from a national
clearing bank, had to be matched pound for pound by the host
educational establishment. The bank's funding gift totalled £100,000
and the pre-condition was met.

Respondent information emphasizes the importance of the existence of
revenue funding for the first two or three years of a project's
life. Receipts from user companies are comparatively low in relation
to running costs, at least in the early years, and the projects
require time to become self supporting financially. While there is a
trend towards charging users for company staff training and consultation services, much individual training is funded by government agency
sources, and project staff see the need to keep consultative service
charges as low as possible, especially to small businesses. There is
some evidence of project uncertainty or ambiguity in relation to
charging users, especially for business advice as, particularly with
the educationally based projects, the project ethos emphasis would
appear to focus on 'service' rather than on financial profit, although
the need for user firm receipts seems not to be underestimated by any
of the case study projects.

Case study information has highlighted both the need for large scale
initial and ongoing capital and revenue funding and the multiplicity
of funding sources which have been used by the projects. Project

staff have additionally emphasized the need for continually seeking
new and different sources of funding for expensive new equipment and
software and additional experienced staff for general expansion
purposes. There would, therefore, seem to be some evidence of a
current and future need for a more unified and systematically cohesive
funding strategy for shared facility projects.

Current operations

Respondent information suggests that while there is considerable
variation within the projects because of the complexity and potential
of the equipment being used, there is a common emphasis on the need to
staff shared facility projects with highly experienced and competent
people. Staffing levels within the five project facilities vary
between three and thirty-two. Respondent information shows that staff
have been, and continue to be, recruited from two main sources;
industry, usually local industry, and local educational establishments. The projects located within educational establishments have
the advantage of the availability of a wide range of technical,
theoretical and practical expertise, much of it on a part time basis.
 All but one of the projects report that they are operating to full,
or near full, capacity in terms of existing staff availability, extent
of office or working space, and availability of equipment. All admit,
however, to somehow finding more of each in order to meet increased
user demand, and all have plans to increase capacity in the future.
 Project staff responses show strong evidence of a high degree of
staff enthusiasm, pleasure and pride in their work, coupled with what
might be termed a 'devoted commitment' to project objectives. Long
hours would appear to be worked by project leaders and some other
staff, some of this unpaid. Excitement and enthusiasm would seem to
compensate adequately for 'unpaid' working time. Project leaders
displayed distinct entrepreneurial dynamism in the engagement of their
work, particularly in relation to locating funding and in planning for
expansion. Project staff excitement would appear to emanate partly
from an awareness of existing high technology applications, but
perhaps more potently from informed speculation as to potential future
developments. Comments like 'the sky's the limit' and 'the possibilities are endless' were recurrent throughout interviews. There is some
evidence of some project staff impatience and disappointment with
colleagues and individuals from industrial or business organizations
who decline to understand or acknowledge the significance of the new
technology available and its direct application both to themselves and
to current industrial development generally.
 There is case study evidence, confirmed by the results of the
preliminary telephone search, which highlights the wide range of
services and equipment offered by the projects, and the constant
development of these services in terms of additions to staff and new
equipment and software packages. All the projects offered a full
range of services including business advice, help in selection of
computer systems and software appropriate to individual user needs,
and in-house and project-located training courses (all with 'hands on'
experience for the student user on a wide range of equipment and
software packages). All the projects had increased their range of
services, staffing levels, equipment and software since inception.
One of the projects aimed its services at a specific product group,
the others aimed more broadly at the general and electrical
engineering sectors of industry.

Case study information shows that project accountability is increasingly a feature of shared high technology facility development. Four of the projects report that they are directly accountable to a governing body or management committee of some kind which incorporates respresentatives mainly from funding bodies, usually local authorities, educational hierarchies and large-scale industry. Three of the projects report that membership of their respective management committee is drawn additionally from organizations in local industry generally. There is therefore evidence of a growing collective interest in the development and utilization of these shared facility projects extending beyond an immediate organizational hierarchy. Only one project, located within an educational establishment, reported that it was accountable to no form of management committee or outside control whatsoever.

While case study information suggests that the projects have managed to develop strong links with a variety of other groups including employers' federations, trades unions, individual companies, funding organizations, government agencies and local and regional authorities, there was one notable exception to a generally encouraging picture of interorganisational interest. Three of the projects reported a distinct lack of interest in their activities from their local Chambers of Commerce, whose attitude seemed to be that high techology developments and applications were not applicable to them or to their members. Only one project reported that it had positive and encouraging links with its local Chamber of Commerce which extended to representation on the project management committee.

Project user groups

The information collected suggests that the small business sector of industry is, and is seen by the projects to be, an important component of shared high technology facility use. While all the projects reported a spread of user groups in terms of firm size, including large national and even multi-national companies, there is evidence to suggest that small businesses represent a substantial proportion of the clientele of shared facility projects. Only one of the projects reported that the majority of its users were big companies; the others all reported a large percentage of small businesses using the facility. Three of the projects estimated that the majority of their users employed between 50 and 200 people, and two projects estimated that 25 per cent of their user firms employed fewer than 30 people. All but one of the projects reported that special efforts were made to attract the small business to the facility, and all the projects felt that the greatest constraint on the small business, in terms of high technology conversion, is a lack of money to invest in expensive equipment.

The results indicate a clearly observable lack of project staff knowledge or experience of women owner-manager or ethnic owner-manager users. Three of the five projects reported very few female or ethnic origin-led firms as clients. Additionally, project staff interviewed, with one notable exception, appeared to have little conception of possible or actual problems which might be experienced by female or ethnic origin owner-managed firms. Given that there is evidence of an increase in women-led small new companies nationally, and given also that the West Midland population is multi-ethnic, it may be significant that the shared facility projects appear not to be attracting women or ethnic origin entrepreneurs to the facilities.

One project had good information on minority group user firms, reporting that a large proportion of users were Asian-led firms, and that female entrepreneurs used the facility, albeit in small numbers. This project reported that women and Asian-led firms face additional problems to those commonly experienced by white, male-led small businesses. Asian businesswomen must contend with additional cultural norms which preclude association with non-family males, and with mobility difficulties, and Asian small businesses tend to be tethered by family commitment and lack of relevant work skills.

It may or may not be significant that this respondent was female (and white). Another project leader, male (and white), revealed a sensitivity to and experience of female-user constraints relating to his facility. It is possible that white, male project leaders and staff in general continue to confine their perception of British business and industry as being fundamentally the domain of the white male. Whether or not this subconscious view underlies project policy and practice, there is evidence to suggest that shared facility projects in general may not be aiming their publicity, or tailoring their services, to women or ethnic minority potential users.

Respondent information was unanimous in declaring that there was no average period definable in terms of user firm support. The strongest factor in determining the period of facility support necessary appears to be product complexity followed by the degree to which the user firm is prepared to accept and accommodate the concept of high technology application. All but one of the projects reported that a significant proportion of users had achieved high technology conversion within the firm, and that more were in process of doing so. The exception was the youngest project, only six months old, which had not yet had time to effect such an influence on client groups. While, in general, project staff saw high technology applications as being an increasingly essential element in current and future industrial efficiency and expansion, case study responses indicate an awareness on the part of project staff that high technology conversion is not necessarily the chief aim of the project as in some instances the solution to a firm's problem may not lie with the adoption of expensive computerized equipment or software. Project staff indicated a concern to provide services which are flexible and appropriate to individual user need, rather than to promote high technology conversion as a universal 'cure all'.

Future expansion and development

Respondent information indicates that the managements of shared facility projects in general realize the importance of publicity and utilize a wide range of facilities in order to advertize project services. All but one of the projects report that strenuous efforts are made to publicize the existence and services offered by the facility, and that policies specifically aimed at attracting the small business are incorporated into their general publicity campaign. Publicity brochures, press coverage, local radio, exhibitions, seminars and open days represent the general means by which the projects advertize their wares. With one exception, the same one, all the projects maintain an annual publicity budget which varies between £5,000 and £25,000. The exception reported that, while they do advertize and hold open days and have stands at some exhibitions, they do not have a policy for attracting the small business, nor do they maintain a publicity budget. Respondent information from this project would suggest a certain impatience at the suggestion that they should need very much self advertisement.

There is strong evidence, from all the case study projects, and from projects covered by the telephone search, of the constant need for monitoring services and upgrading equipment and software so as to meet changing client group need. All the projects questioned emphasized the need for flexibility and responsiveness to new technological development and to individual client firm need. There is some evidence from project staff response that the focus of industrial development has swung from being industry led to being education or project led, which implies a project staff responsibility to lead industry safely through the complex labyrinth of the new technology.

Respondent information identifies firm evidence of a strong conviction on the part of project staff that the new technology has applications to all sectors of industry and commerce and that it is the key to future industrial efficiency and competitiveness. Preliminary telephone search information confirms the reponses of the case study projects about the potential for shared high technology facility applications throughout all sectors of industry, in that shared facilities are in existence, or are planned, for the metal-based industries and the car component industry. Shared facilities exist for precision calibrating and measuring-gear, for calculation of output and monitoring of manufacturing machinery, and for new product and new processes generally. Case study information suggests that the leather industry, jewellery and fashion designers, furniture manufacturers and the soft furnishing trade could all benefit from shared high technology facilities, as could individuals pursuing a personal hobby or interest. All projects reported that the major benefits of conversion would be firstly, better control of manufacturing processes and resources, followed by better quality products and jobs. There was general consensus that shared facilities should be more readily available to industry, particularly small-sale industry and business, because of the expense involved in high technology acquisition, and because of the complex nature of the multiplicity of equipment and the software on the market.

There is a clear trend among the projects to look ahead and develop new and better services for users. While project responses deny meaningful constraints on project expansion, consideration needs to be given continually to the availability of the high cost of equipment and software packages, and both the availability and the cost of experienced staff. One case study project is currently involved in imminent plans to move location to larger and purpose-built accommodation. As well as a commitment to facility development, project staff show considerable ingenuity and drive towards developing personal interests connected with their work. One project leader is considering taking on the establishment and organization of a regional marketing consortium, while another is interested in involving herself in the establishment of female-led cooperatives in the locality.

User firms survey

Case selection

Three of the case study projects were willing to give the names of user firms from which a small-scale random survey of project user firms was conducted by telephone. A short questionnaire was devised which asked for basic information on the size of the firm and the nature of its operations, and which sought details on the nature of the firm's relationship both with the project facility itself and with user firm conversion to high technology manufacturing systems.

The user firms - a general description

Five user firms were surveyed. All are located in the West Midlands region and all are manufacturing companies, four in the general and electrical engineering sectors of industry and one in the textile and clothing sector. All five firms have converted to high technology computerized manufacturing and/or design systems. Two of the firms are part of a larger group of companies; three are independent manufacturing units. There is a wide variation in the size of the firms surveyed. The firms employ 56, 500, 1,050, 1,800, and 2,500 people respectively. All five firms were most responsive and cooperative with this research.

Findings from user firms survey

There is some evidence to suggest that the element of chance may play a role in the initial contact between the user firm and the project facility. No overall pattern of contact emerged from user firm information on how they came to hear about, and subsequently to contact, the project facilities. Two of the user firms were locational neighbours of two of the projects and had come to know of their existence by this means. Another user firm 'came across the project facility' during its research into high technology conversion of the firm's manufacturing processes. Another firm made contact after receiving a mailshot sent by the project, and the fifth firm had read a press release.
User firm information suggests that the two project facility services most widely used are likely to be staff training and advice on the appropriate choice of software packages. All five user firms used the project facilities for training on computers, either for administrative, manufacturing, or design purposes, and in four cases for all three. Three of the user firms report using the project facilities for advice on choice of appropriate hardware and software options for their individual firm, and a fourth regrets that it did not do so at the time of its high technology conversion and states that it will do so in future. A wide range of project facility services were utilized by the user firms. Two of the user firms report using facility services to design in-house training packages, and three of the firms required training on word processing and training on management and office administration packages. Three of the projects required very specific training for staff involved in computerized design processes, production management and manufacturing processes. One firm used the project facility to recruit trained staff and another rented accommodation for in-house staff training. The wide range of services utilized by the user firms would tend to confirm project staff awareness of the need to constantly monitor its services and to preserve its flexibility of response.
Survey responses provide strong evidence of user firm satisfaction with project facility services and attitude. All five user firms stated that they had experienced no problems in their relationship with the project facilities. A special mention was made by two of the firms regarding the willingness of the respective projects to meet the

specific needs of the firm, both in terms of training content and timing of courses. One firm had a slight reservation about the relevance of one particular course but stated that the firm itself was partly to blame in this instance in not stating its needs more clearly.

There is evidence to suggest that initial staff anxiety and resistance is the major problem to be faced by firms wishing to undergo high technology conversion. Four of the user firms specified this as the major problem they had been forced to overcome. Only one firm mentioned time constraints as being a problem and none of the firms mentioned that they had experienced financial constraints in this respect. Three of the firms mentioned how important it was to prepare fully to reassure staff well in advance of the actual conversion time, and to offer them competent and complete training on the new technology.

There is evidence to suggest that one of the most valuable services which project facilities can offer to their user firms is that of advice on the selection of appropriate hardware and software for individual user firm need. Four of the user firms regard the chief benefit accruing to them from their relationship with the project facility to be the adoption of the most appropriate computer systems for their need. In support of this, four of the user firms also reported a major benefit of their contact with the projects as being staff knowledge of computers, computerized systems and the range of software packages available to the firm.

A subsidiary finding from the survey evidence is that user firms regard the chief benefits accruing from successful adoption of appropriate high technology as being first, better control of production and manufacturing processes and resources; second, better quality products produced with greater precision; third, better relationships with customers and suppliers due to better production and stock information; and fourth, improved scope for design staff.

8 Conclusions and policy implications

Conclusions and policy implications

Growing evidence supports the hypothesis that a relative failure to innovate in the UK has been a significant contributory factor in the declining competitive position of UK firms, particularly in certain key (often manufacturing) sectors and regions of the country (such as the East and West Midlands). Many commentators look to current small firms as the source of regeneration and future prosperity. However, while there has been a growth in the size of the small firm sector, there is an increasing concern over the ability of small firms to undertake technological and organizational innovation, and thereby achieve growth and large firm status.

Just under half the Leicester firms surveyed use some form of computerized office administration system such as word processing, computerized wages, invoicing, payments, stock and/or production control systems. Fewer than a third of the firms use any form of computerized manufacturing system and only a small minority have a computer aided design system. Firms who are not computerized in any area have not, in significant numbers, even considered using a computerized system. The pattern of use suggests that a computerized administrative system is the most likely to be adopted and could normally be expected to precede any other form of computerization (though one firm had a CAD system and does not use any form of computerized office system).

Case study interviewees suggested that the main benefits of computerization of production were the opening up of new markets, the improvement in quality of the finished product and the speed and flexibility offered by the new machinery. They suggested that the main disadvantage of using computerized systems was the resultant overdependence on outside expertise. Those companies who had adopted did however feel that the systems had been worthwhile and had lived up to their expectations, though many had found hidden costs such as the inability of the equipment to cope with low quality materials. In terms of office administration systems, firms reported problems with manufacturers' back-up services and overenthusiastic claims by sales representatives of the suitability of hardware and software for the firms' specific purposes.

Although we did not ask any questions relating to the potential for local authority involvement (other than relating to a shared technology and training facility) in the survey of firms, we pursued this in the Leicester case study interviews. Firms on the whole were reluctant to make suggestions for local authority intervention and several felt it inappropriate that the authority should intervene. Clearly there may be an underlying fear both with the tax rate implications of such intervention and with the power and autonomy of the small firm. However, in addition to the suggestions made in relation to a shared technology facility referred to above, the following suggestions were made: (i) promotion of the industry to improve its image; and (ii) promotion abroad of the industry and the city. These two suggestions gained a majority of support from the case study companies. Such measures would not only help to increase the demand for local products, but also help in the recruitment and maintenance of a skilled and talented work force. Exactly half of the companies suggested a reduction in the rates. There were, however, a number of other areas where firms appeared to require assistance. These are dealt with in detail elsewhere (Bosworth, Jacobs and Lewis, 1988), but include the following: the provision of relevant information in an easily digestible form; loan facilities to help with cash flow problems at a time when firms want to make the crucial size transition (often after four or five years of trading); the provision of suitable premises in the most appropriate locations; the availability of well designed, coordinated and high quality training courses; representation at trade fairs (particularly foreign) when individual managers from smaller firms could not spare the time or afford to go.

Our analysis of the survey findings suggests that the major factors influencing the adoption of new technological systems are the size of firms, with larger firms being much more likely to adopt than small firms (100 employees being a critical size), and whether or not firms supply chain stores, those firms so doing being much more likely to have adopted than otherwise. This would suggest that the chain stores exert a certain amount of pressure on their supplying firms to conform to quality and delivery times which puts pressure on firms to operate using the best-practice technology. Knitwear firms were found to be more likely to have adopted new technological systems than other firms, reflecting the advances which have developed in knitwear design and production machinery as well as the demands of quality, delivery, and colour matching requirements in the sector.

Although there has been a phenomenal increase in the number of start-ups in the clothing and textile industry as well as other sectors, and while there has been growth in the size of the small firm sector, there is concern that the small firm, if it survives, will remain small and thereby fail to generate income and employment growth. The possible reasons for this failure are traced to the explanation of the conundrum associated with why low technology/low skill firms find it possible to coexist alongside high technology/ high skill firms. The answer lies in the fact that relative factor prices tend to result in a selection procedure that leads to small firms being, on balance, relatively low technology and low skill. As a consequence, given that the ability to successfully adapt and adopt new technologies and organizational forms are related to the quality of the management and the work force, major competitive shocks which reduce prices tend to force small firms to respond by searching for lower cost inputs and this results in a vicious circle of reduced input quality and even lower competence in innovation. The question then arises about the way in which this downward spiral can be broken. One possible method is the provision of shared facilities that can increase the supply of knowhow to firms in need and, in a variety of ways, reduce the costs of the low to high technology transition.

Around half of the firms in the Leicester textiles and clothing survey would consider using a shared facility which offered the use/demonstration/advice of computerized systems and a similar proportion would consider using a training facility. This is by no means an overwhelming response given the nature of the offer. Such facilities have, however, been introduced elsewhere (e.g. Birmingham) where they have proved successful and their use in engineering industries has been widespread, again with much success. In view of case study comments in respect of computerized administrative systems and also because it has been shown that office administration is usually computerized before design and manufacture, it would seem logical to consider facilities in this area as a priority over design and manufacture.

The case study interviewees suggested the following features, in order of priority, of any shared facility which might be introduced in Leicester: 'hands on' experience of computerized equipment; information and advice on available technology which was company specific, neutral, expert and backed by experience of the industry; similar advice relating to technology appropriate to the firm's needs; and finally, a short term rental service for use on company premises.

There is evidence from other countries that there has been a more systematic development of the infrastructure, including a variety of shared facilities such as the Emilia Romagna region of Italy. Despite the rather negative attitude of the Leicester-based firms, the empirical work undertaken as part of the ESRC study reveals a rapid growth in the provision of shared facilities in the UK and an equally rapid uptake of the services on offer by user firms. The evidence appears to indicate that at least to date this has resulted in the successful adoption of high technology systems by client firms and, in turn, this has led to improved performance among user firms. Perhaps in this instance the supply of such facilities effectively creates its own demand.

While this is a fairly positive picture a number of worries emerged from the case studies of both shared facility projects and user firms. The first of these was the almost capricious nature of the initial funding. The second was the need of the various projects to achieve a fairly rapid transition to self sufficiency in terms of finance which led to a conflict in goals which may have influenced the composition of their client groups. By implication, the third problem area concerns the composition of the user groups and, in particular, the question whether truly small firms (i.e. those with fewer than 50 employees) were underrepresented, the lack of use by minority groups (i.e. female and ethnic minority owner managed firms) and, finally, the danger that the shared facilities were used (and often successfully) by those best able to use them rather than by those in most need. The latter results may reflect both the more immediate patterns of demand and the need for the shared facility project to be

seen to succeed. It is an important question whether the need to be
seen to succeed, coupled with the need to achieve self sufficiency,
can ever allow these types of projects to wholly meet the needs of the
more problematic potential users given the way they are currently
funded. There appears to be a number of direct analogies between
shared facilities and small firms, with the success of both dependent
on funding arrangements and the drive and entrepreneurship of the
individuals who run them.

Bibliography

Apparel International (1986 and 1987), London Piel Caru Publishing Ltd.

Armstrong, K.M., Jennett, N. and Lewis, J.A. (1985), 'Report on the Engineers' Small Tool and Gauge Sector in the West Midlands and the United Kingdom' to the Economic Development Unit, West Midlands County Council; Wolverhampton Economic Research Unit, The Polytechnic, Wolverhampton.

Armstrong, K.M. and J. Lewis (1986), 'Skill Shortages in the West Midlands Engineering Industry', Final Report, Wolverhampton Economic Research Unit, The Polytechnic, Wolverhampton.

Binks, M. and Vale, P. (1984), 'Finance for the New Firm', NUSFU Paper No. 2, University of Nottingham.

Birnhaun, B. et al. (1981), 'The Clothing Industry in Tower Hamlets', Tower Hamlets Council.

Boggon, B. (1987), Secretary, Leicester and District Knitting Industry Association Ltd. (in conversation).

Bosworth, D.L. (1987), 'Barriers to Growth in Small Innovatory Firms: Labour Market Factors', Paper presented at the ACARD Seminar on Barriers to Growth in Innovatory Small Firms, Rolls Royce, February, London, (Discussion Paper No. 36, Institute for Employment Research, University of Warwick, Coventry).

Bosworth, D.L. and Jacobs, C. (1987), 'Barriers to Growth in Small Innovatory Firms; Management Attitudes, Behaviour and Abilities'. Paper presented at the ACARD Seminar on Barriers to Growth in Innovatory Small Firms, Rolls Royce, February, London, (Discussion Paper No. 37, Institute for Employment Research, University of Warwick, Coventry).

Bosworth, D.L., Jacobs, C. and Lewis, J. (1987), Shared Facilities, Human Capital and Growth in Innovative Small Firms, Report to ESRC.

Bosworth, D.L., Jacobs, C. and Lewis, J. (1988), The Textile and Clothing Industry in Leicester, Project Report, Institute for Employment Research, University of Warwick, Coventry.

Bosworth, D.L. and Wilson, R.A. (1987), <u>Infrastructure for Technological Change</u>, Research Report, Institute for Employment Research, University of Warwick, Coventry.

British Textile Confederation (1987), in Knitting International, vol. 94, p. 1121, June.

Burgess, C. (1985), 'Skill Implications of New Technology', <u>Employment Gazette</u>, October.

Carter, C. (ed) (1981), 'Industrial Policy and Innovation'. Heinemannn Educational Books.

Cole, R., (Chair) NEDO Steering Group (1987), 'Training in the Clothing Industry' in Apparel International, vol. 12, no. 2.

Dasgupta, P. and Stiglitz, J. (1980a), 'Uncertainty, Industrial Structure and the Speed of R&D', <u>Bell Journal</u>, Spring, pp. 1-28.

Dasgupta, P. and Stiglitz, J. (1980b), 'Industrial Structure and the Nature of Innovative Activity', <u>Economic Journal</u>, vol. 90.

EITB (1983), 'The Technician in Engineering', Research Report no. 9.

Francis, <u>et al</u> (1982), 'Microelectronic-based Production Technologies', Department of Social and Economic Studies, Imperial College, London.

Grabowski, H. G. and Mueller, D. (1987), 'Industrial Research and Development, Intangible Capital Stocks and Firm Profit Rates', <u>Bell Journal</u>, vol. 9.

Green, A.H. (1987), British Narrow Fabrics Association (in conversation).

Griliches, Z. (1957), 'Hybrid Corn and Exploration in the Economics of Technological Change', <u>Econometrica</u>, vol. 25, pp. 501-22.

Griliches, Z. (1958), 'Research Costs and Social Returns: Hybrid Corn and Related Innovations', <u>Journal of Political Economy</u>, October.

Griliches, Z. (1960), 'Hybrid Corn and the Economics of Information', <u>Science</u>, 29 July, pp. 275-80.

Griliches, Z. (1980), 'Hybrid Corn Revisited: a Reply', <u>Econometrica</u>, vol. 48, pp. 1463-6.

Hall, G. (1987), 'Lack of Finance as a Constraint on the Expansion of Innovatory Small Firms', Paper presented at the ACARD Seminar on <u>Barriers to Growth in Innovatory Small Firms</u>, Rolls Royce, February, London.

Harrison, P. (1987), 'ITMA 1987', in <u>Textile Horizons</u>, vol. 7, no. 11, The Textile Institute, Manchester.

Institute for Employment Research (1987), <u>Review of the Economy and Employment</u>, University of Warwick, Coventry.

Jacobs, C., Lewis J. and Roberts, M. (1987), 'A Local Authority Strategy for Intervention in Clothing and Textiles', Paper presented to a seminar - Changing Technology in Textiles - organised by LATC (Local Action for Textiles and Clothing), Bradford.

Jennett, N. (1984), 'Report on the Pumps, Valves and Compressors industry in the West Midlands and the United Kingdom' to the Economic Development Unit, West Midlands County Council; Wolverhampton Economic Research Unit, the Polytechnic, Wolverhampton.

Johansen, L. (1972), <u>Production Functions: Short Run and Long Run, Micro and Macro Aspects</u>, North Holland, Amsterdam.

Johansson, B. and Karlsson, C. (1985), 'Industrial Applications of Information Technology: Speed of Introduction and Labour Force Competence', Working Paper 1985:4, CERUM, Umea Universty, Umea.

Knitting International (1986 and 1987), Ferry Pickering Publishers Ltd., Leicester.

Knitting Sector Working Party (1983), 'The British Knitting Industry: Prospects and Profits in the 1980s', National Economic Development Office, London.

Koutsoyiannis, A. (1975), Modern Microeconomics, Macmillan, London.

Leach, H. (1987), 'British Textile Confederation President', in Review of Textile Year 1986, Knitting International, vol. 94, p. 1122.

Leigh, R. and North, D. (1983), 'The Clothing Sector in the West Midlands', Report to the Economic Development Unit of West Midlands County Council; Wolverhampton Economic Research Unit, the Polytechnic, Wolverhampton.

Lewis, J.A. (1983), 'Report on the Fastener Industry in the West Midlands and the United Kingdom', to the Economic Development Unit, West Midlands County Council; Wolverhampton Economic Research Unit, the Polytechnic, Wolverhampton.

Lewis, J.A. and Armstrong, K.M. (1986), 'Skill Shortages and Recruitment Problems in West Midlands Engineering Industry', National Westminster Bank Review, November.

Lindley, R.M. and Wilson, R.A. (eds) (1989), Review of the Economy and Employment 1988/89 - Volume 2: Occupational Studies, Institute for Employment Research, University of Warwick, Coventry.

Lipson, E. (1953), 'A Short History of Wool and its Manufacture', Heinemann, London.

Littler C. and Salaman, G. (1984), 'Class at Work: The Design and Control of Jobs', Batsford.

Loman, P. (1987), National Union of Hosiery and Knitwear Workers (in conversation).

Marris, R. (1963), 'A Model of the Managerial Enterprise', Quarterly Journal of Economics.

Marris, R. (1964), The Economic Theory of Managerial Capitalism, Macmillan, London.

Metcalf, S. (1987), 'Barriers to the Growth of Small, Innovatory Firms', Report to the Advisory Council for Applied Research and Development, Provisional Draft, March.

Millington, J.T. (ed) (1987), in Knitting International (in conversation).

Millington, J.T. (ed) (1987), in Knitting International vol. 94, pp. 1118, 1119 and 1124.

Mueller, D. C. (1967), 'The Firm Decision Process: An Econometric Investigation', Quarterly Journal of Economics, vol. 81, pp. 58-87.

Nathan, C. (1986), Apparel International, vol. 10, no. 6.

NEDO (1983), 'Skills and Technology: the Framework for Change', Heavy Electrical Machinery EDC.

NEDO (1987), 'Dynamic Response', Knitting EDC report, NEDO.

Northcott, J. and Rogers, P. (1984), Microelectronics in British Industry: the Pattern of Change, PSI No. 625, March.

Northcott, J. and Walling, A. (1988), 'The Impact of Microelectronics: Diffusion, Benefits and Problems in British Industry', PSI No. 673.

Parkinson, M. (1987), Joint Head of the School of Textile Technology, Leicester Polytechnic (in conversation).

Pindyck R.S. and Rubinfeld, D.L. (1976), Econometric Models and Economic Forecasts, McGraw Hill.

Robson, R. (1957), 'The Cotton Industry in Britain', MacMillan, London.

Samuels, J. S. and Smyth, D. (1968). 'Profits, Variability of Profits and Size of the Firm', Economica, May.

Senker, P., Swords-Isherwood, N., Brady, T. and Huggett, C. (1981), 'Maintenance Skills in the Engineering Industry: The Influence of Technological Change', EITB Occasional Paper No. 8.

Singh, A. and Whittington, G. (1968), <u>Growth, Profitability and Valuation</u>, Cambridge University Press, Cambridge.

Solo, R. (1966), 'The Capacity to Assimilate an Advanced Technology', <u>American Economic Review; Papers and Proceedings</u>, May. pp. 91-7.

Steer, P. and Cable, J. (1977), 'Internal Organisation and Profit: an Empirical Analysis of Large UK Companies', Discussion Paper 77-47, International Institute of Management, Berlin.

Swan, P. (1957), 'Abbey Hosiery Mills at ITMA 87 Review', The Textile Institute, Midlands Section Annual Symposium.

Teague, E. (1987), Software Systems UK Ltd., Oldham, Lancashire (in conversation).

<u>Textile Horizons</u> (1986 and 1987), vol. 6, no. 12 and vol. 7, no. 1. The Textile Institute, Manchester.

Textiles Statistics Bureau (1979, 1980, 1981, 1982, 1983, 1984 and 1986), Pocket Facts about the UK Textile Industry, Manchester.

Todd, D. (1971), <u>The Relative Efficiency of Small and Large Firms</u>, Committee of Inquiry on Small Firms, Research Report No. 18, HMSO, London.

Totterdill, P. and Pearce, S. (1986), 'Prospects for Local Authority Intervention in the Textiles and Clothing Industry', Centre for Local Economic Strategies (CLES), Research Report no. 4.

Walker, A. (1984), 'Report on the Locks, Keys etc. industry in the West Midlands and the United Kingdom' to the Economic Development Unit, West Midlands County Council; Wolverhampton Economic Research Unit, the Polytechnic, Wolverhampton.

WERU Sector Studies, Reports to the Economic Development Unit, West Midlands County Council.

Wheatley, R. (1987), Joint Head of the School of Textile Technology, Leicester Polytechnic (in conversation).

Wildsmith, J.R. (1984), 'Report on the Metal Windows and Doors industry in the West Midlands and the United Kingdom'', to the Economic Development Unit, West Midlands County Council; Wolverhampton Economic Research Unit, the Polytechnic, Wolverhampton.

Williams, I. (1988), 'Steel Polishes its Image', <u>Sunday Times</u>, 30th October.

Williams, V. (1984), 'Employment Implications of New Technology', <u>Employment Gazette</u>, May.

Wilson, R. and Bosworth, D.L. (1987), <u>New Forms and New Areas of Employment Growth</u>, Programme of Research and Actions on the Development of the Labour Market, Draft Final Report of the UK Study CEC Study No. 85397, Institute for Employment Research, University of Warwick, Coventry.

Wilson, R.A. (1989), 'Occupational Assessment' in Lindley, R.M. and Wilson, R.A. (eds) (1989), <u>Review of the Economy and Employment 1988/89 - Volume 1</u>, Institute for Employment Research, University of Warwick, Coventry.

Wilson Committee (1979), 'The Financing of Small Firms: Interim Report of the Committee to Review the Functioning of Financing Institutions', Cmnd 7503, HMSO, London.